Contents

KU-512-807

Preface

Since 1945 there have been many advances in geomorphology. It is perhaps no overstatement to say that the changes which have taken place amount to little less than a revolution. It is not only a matter of rapidly increasing knowledge and the introduction of new techniques, the approach of the modern geomorphologist differs in many respects fundamentally from that of a generation ago. Some of the new concepts are highly sophisticated, and published material tends to be specialized. The reader is thus faced with a multiplicity of sources, many of which are complex and in which the author assumes an understanding of the required mathematical expertise. The aim of this book is to present in one volume the more important recent advances in geomorphology, in such a way that the student is able to work with a minimum of tutorial assistance, and without the assumption of other than the most elementary mathematical knowledge. Where suitable sources for further reading are readily available, they are referred to in the text.

Research in geomorphology has proceeded over a very wide field, and the selection of what to include in a relatively short book is a matter of judgement. There can be no attempt to be comprehensive. Ideas and techniques are included which are at an appropriate level for students and which may be:

1. necessary for an understanding of modern theory;
2. difficult, and inadequately explained in published texts for those without the necessary (sometimes mathematical) background in that field;
3. taken from sources not readily available;
4. useful in practical work, including fieldwork.

Every chapter will contain a 'consolidation' section, consisting of questions and exercises, to ensure that the content of the chapter has been fully understood. Not all of the answers will necessarily arise directly out of the text. To some questions there *are* no generally agreed answers, only differences of opinion. These will require some personal thought and investigation. They are posed in an attempt to encourage the student to examine facts critically and encourage personal investigation, by showing him that not all problems in geomorphology can be resolved by consulting the appropriate text, but require further research. Sometimes solutions are not yet available, and all we have are suggestions and differing hypotheses. It is hoped that this method will help him in the difficult process of learning to think for himself.

Chapter 1

Introduction

Early landscape models

The cycle of erosion model
Before 1945 the method of landscape evaluation in terms of structure, process, and stage (i.e. time) proposed by W.M. Davis, and systematized in his model of the normal cycle of erosion, remained comparatively unchallenged in Britain. (It is true that in the 1920s W. Penck cast some doubt on the assumptions made by Davis, and proposed alternative ways of slope development that Davis had not considered up to that time; although the ideas of Penck did not have great impact initially, more recently they have been considered very seriously.) Most geomorphologists now have serious doubts about the usefulness of the cycle concept as a way of studying landforms. It is not possible in the space available to describe the erosion cycle nor Penck's contribution to the theory of slope evolution in any great detail, but a brief outline is necessary in order to understand the argument which follows. (A useful treatment of these topics will be found in Small, 1970, and Sparks, 1973.)

Davis, working mainly in the temperate lands of the U.S.A. and Europe, studied landscape **evolution**. He postulated (for the sake of simplicity) a flat area of land rapidly uplifted. He believed that the landforms which subsequently developed upon it were a function of the rock type and structure, the kind of erosion to which it was subjected (process), and the length of time the agents of erosion

Fig. 1.1 *The Davisian cycle: the evolution of landscape through time*

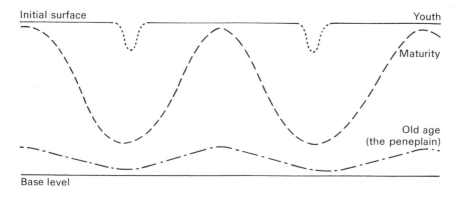

Initial surface   Youth

                                                                    Maturity

                                                        Old age
                                                        (the peneplain)

Base level

had been acting upon it (stage). He described the stages of evolution through which the landscape passed as those of youth, maturity, and old age. The stage of youth lasted while the land was still high and much of the original surface remained; valleys were deep and steep-sided, with actively eroding rivers. Then came maturity, with the landscape beginning to wear down, valleys becoming wider with rounded interfluves and slopes becoming less steep. There was a period of maximum relief when the widening valley left only the very crests of the hills as part of the original surface. Finally came the end of the cycle with the low, gently inclined landscape of old age, in which sluggish rivers flowed across flattish floodplains, and erosion and deposition were roughly in balance (Fig. 1.1). Essentially the concept was one in which the uplifted land was progressively worn down until the main transporting agents (the rivers) approached base level. These steadily lost their capacity to erode vertically, until at last a nearly featureless peneplain resulted.

The cycle of erosion model assumes continuous progression from youth to old age (with interruptions allowing for intermediate changes in base level within the cycle, e.g. rejuvenation). However, Davis also recognized that some rivers, or parts of rivers, appeared to be in a state of equilibrium, in which denudation of the hillslopes balanced the transporting power of the rivers. He was therefore obliged to introduce into the cycle the concept of grade and the graded profile.

The contribution of W. Penck

Penck, working in South America, realized that not all landforms could be classified in terms of the Davisian cycle. In landscapes that could otherwise be classified as mature he saw slopes that were constant or rectilinear (i.e. straight in the sense that the angle of slope remains unchanged), when theory predicted they ought to have been rounded. He suggested that Davis's simplification in assuming rapid uplift was misleading, because there is evidence that tectonic forces do not operate in this way. He believed that landforms depend upon the relative rate of uplift of the land and the efficiency of the forces of removal acting upon it. The agent relating these two forces was the river, which controls slope development by the rate at which it cuts down into the valley floor.

During a period of accelerating uplift when streams are downcutting energetically, and relief is increasing, he considered that valley slopes will tend towards convexity. If the rate of downcutting declines, streams will be progressively less able to transport material provided by downslope movement, deposition will occur, and valley sides will tend first to be straight and then develop concave forms. If erosion by the river is constant, then a state of equilibrium may exist, and the valley sides may remain rectilinear. Penck's idea of the evolution of landforms therefore was independent of an erosion cycle, but rather related to the energy of rivers in relation to the uplift of the land. Increase or decrease in river energy could be controlled by a number of factors. These might be changes in the rate of uplift, in base level, in runoff, in long-term variations in climate, or in a combination of these.

Probably Penck's most important contribution to geomorphology was in his study of slope development. It has often been stated that, because he suggested a rectilinear slope element would, as a result of denudation, retreat parallel to itself, valley sides as a whole would retreat in parallel. From this came the idea that whereas Davis contended landforms were wearing down, Penck believed they were wearing back. Although a widely held view, this is not strictly the case. Penck's theory of how some of the slope profiles he observed might originate is summarized in Fig. 1.2.

Penck believed that the rate of removal of weathered material from a slope depended on steepness rather than the process involved, so that a steep slope will retreat more rapidly than a gentle one. Also, if weathering takes place equally over all the surface of the slope, the removal of the weathered product will cause the slope face to retreat parallel to itself. Fig. 1.2a illustrates this theory: AB is assumed to have been the initial steep profile of a valley side; as weathered material is removed from the surface of the slope, 1, 2, and 3 are subsequent positions of the profile after three successive time intervals have elapsed. Of course, weathering and removal is a continuous process which cannot be shown in the diagram. In reality the upper portion of AB would have retreated slowly and continuously until it arrived in the new position CD, parallel to, but further back from, its initial position. The valley would thus grow in width.

Fig. 1.2 *W. Penck's idea of how slopes might develop in homogeneous material. Unimpeded removal, but no erosion, is assumed at the base of the slope. For explanation see text.*
(a) *Retreat of the initial slope*
(b) *Subsequent development*
(c) *Evolution of the cross-profiles of an area between two river valleys*

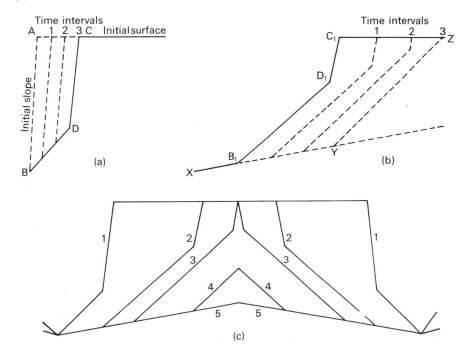

It will be seen that a new slope element, BD, has been created with the retreat of the initial valley side. It is assumed that BD is formed in the following way. The material created by the weathering of the initial surface would move rapidly down to the bottom of the slope. Here there is assumed to be a transporting agent (e.g. a river) which will remove any material deposited in it, but which will not cause any downcutting. We therefore have unimpeded removal with no alteration in the base level of the slope. But as the material from above moves over the bottom part of the slope, the latter is protected from the agents of weathering. The effect is to create a new slope at a less steep angle growing upwards from the base and beginning to consume the steeper slope above it.

Look at Fig. 1.2a and at the profile after the lapse of three time intervals. Imagine small weathered fragments from CD falling, or being washed down on to BD. The angle of BD is less steep, and so the movement of material downwards under gravity will correspondingly slower. The surface of BD will thus be covered with a carpet of weathered products protecting the underlying rock to a greater or lesser extent from further weathering. As denudation continues so the lower, sometimes called basal slope will continue to grow and the upper slope element, as a proportion of the whole profile, be reduced.

The course of further possible slope development according to Penck is shown in Fig. 1.2b. The short steep upper slope is now represented by $C_1 D_1$, and the lower slope by $B_1 D_1$. It will be seen that after a further three time intervals have elapsed, the whole of $C_1 D_1$ will have been consumed by $B_1 D_1$ in its new position at YZ.

While these developments are taking place on the upper slopes, the lower part of BD, having been formed first, will itself have sustained a very long period of weathering and denudation. It is reasonable to assume, as in the case of the initial slope, that material moving downward will tend to protect the bottom part of the slope from denudation. The result will again (and for the same reasons) be the emergence of a new slope element of yet gentler gradient, in this case XB_1. After the lapse of three time intervals the profile of the valley side is represented by XYZ.

As gradients become less steep and the transport of material over the slopes becomes progressively slower, the time taken for the evolution of new slope elements, such as XY, becomes very long indeed. Nevertheless, if other conditions remain unchanged it is to be assumed that the whole slope will eventually be reduced to the gradient of XY right up to the watershed.

It will be seen that after a long period of erosion, the resulting landform (5 in Fig. 1.2c) is not unlike a Davisian peneplain, and that this wearing down of high land has been achieved through a process of the wearing back of individual rectilinear slopes. Some real landforms may be found that very closely resemble the five theoretical profiles depicted in Fig. 1.2c. However, Penck also recognized that much more complex forms were often to be found in nature. He suggested that most slopes, even those which appear at first sight to be curved, really consist of a complicated series of small straight elements, each element retreating parallel to itself. (Slope problems are dealt with further in Chapters 4, 5, and 6.)

Difficulties associated with the theories of Davis and Penck

The cycle of erosion is an example of **deductive** reasoning. That is, argument based on a certain assumption, or assumptions; in this case firstly that an area is rapidly uplifted, and secondly that landforms are a product of continuous evolution (from the forms of youth to those of old age) and depend on structure, the kind of denudation processes involved, and the length of time these have been in operation. This method of investigation is valid only if the assumptions on which it is based are valid. (It may be compared with the more scientific **inductive** method of reasoning, which builds on experimental and quantitative evidence gathered from the laboratory or measured in the field.) There is good reason to believe that rapid uplift on a large scale does not occur, but that it begins slowly and increases (with fluctuations) to a maximum, and then dies away. Hence the fundamental association in Penck's mind of the relationship between rate of uplift and landforms.

There is evidence to suggest that some landscapes appear to be in a state of equilibrium (discussed in Chapter 4) in which the amount of material passing downslope into a river is fairly precisely adjusted to the load the river is able to transport, thus inhibiting further downcutting. This was acknowledged by Davis with the recognition that the long profile of a river could become graded. Yet it is evident that a landscape of equilibrium cannot logically be fitted into a theory which demands continual evolution to an ultimate stage. This was a contradiction never resolved by Davis.

Although as a result of recent research and quantitative measurement tenable hypotheses are beginning to emerge, insufficient evidence has so far been accumulated to permit the formulation of precise laws concerning landform evolution.

Because the continued process of denudation necessarily removes much of the evidence of earlier stages there is little evidence to show that landscapes do evolve in the ways suggested, and none that steeper slopes are invariably younger than more gentle ones.

The cycle model has probably been perpetuated for so long partly because it had become firmly enshrined in many texts, and partly because it is such a simple and effective vehicle for teaching.

Time and space

In 1965 an important contribution to the development of landform as a function of time and space (area) was made by Schumm and Lichty. They argued that the kind of model we construct for the study of landform development depends upon the length of the time-span we have in mind. There is no doubt that if there is no uplift the land is slowly worn away and the surface continually lowered towards base level. If we consider the evolution of landscape over very long periods of geological time the cycle of erosion model may be appropriate. In a period of time as long as this, the violent fluctuations of base (sea) level during the recent Pleistocene would appear as a relatively short aberration in the long process of erosion. (A measure of just how unusual present conditions are is that we have to go back some 250 million years to the late Carboniferous before we find evidence of a previous ice age.) Over such a long period time is the most important independent variable; climate (process) and geological structure become progressively less important as the cycle proceeds. In fact, we are living at a time when there have been very great changes in sea level in recent (historical and archeological) times, with accompanying changes in mid-latitude temperatures and precipitation. The result has been considerable and rapid modification of landforms, through sculpture by ice, glacial, and fluvio-glacial deposition, and variations in base level and fluvial activity.

Schumm and Lichty suggest that the erosion cycle can only be regarded as an adequate model over very long periods: **cyclic time**, in which landforms slowly lose energy and mass as the agents of denudation reduce altitude. However, if the time period over which we are looking is greatly reduced, some streams or parts of streams in a drainage network and the slopes above them may be regarded as at grade, or in a state of dynamic equilibrium. This time period is termed **graded time**. During a period of graded time minor fluctuation may occur as a result, for example, of cyclic variations in precipitation causing consequent cyclic changes in stream flow and sediment discharge, and thus minor alterations in long profile gradients. The landscape would be slowly changing, but due to self-regulating mechanisms in the process involved (discussed in Chapter 3) the drainage network may be regarded as being in a state of equilibrium. If the study is over a still shorter time a steady state may be found to exist, in which erosion, transport, and deposition exactly balance. This Schumm and Lichty term **steady time**. Under these conditions stream long profile gradients would remain unaltered. Fig. 1.3 shows the relationship between cyclic, graded, and steady time.

It is also necessary to appreciate the importance of space (or area). A small drainage system may consist of a single stream; a large system may have within it many component drainage areas. At any time during a cycle one or more small component areas may be graded; that is, it will be in a state of equilibrium, while other parts of the system are not. To achieve a condition of steady state, the area involved would normally tend to be smaller still, and might consist of part of a stream, or an individual hillslope.

Fig. 1.3 (a) *The progressive reduction of long profile gradient as the cycle proceeds. The tiny part of cyclic time represented by graded time is indicated by the thickness of the line*
(b) *Over the very much shorter period of graded time the long profile gradient may fluctuate about a mean, but the variations are never great. Over a still shorter period of steady time the gradient remains unaltered*

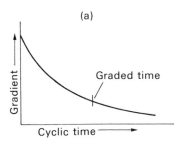
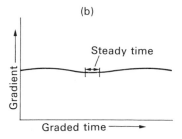

The authors recognize ten drainage basin variables:

1. Time	7. Hydrology (runoff and sediment
2. Initial relief	yield per unit area)
3. Geology	8. Drainage network morphology
4. Climate	9. Hillslope morphology
5. Vegetation	10. Hydrology (discharge of water and
6. Relief (volume above base level)	sediment from system)

For cyclic time, variables 1 to 4 are all independent, with time the most important. The remaining variables are dependent upon these four. Over the very much shorter span of graded time, variables 5 to 7 may also be regarded as independent. Only variables 8 to 10 are dependent as they adjust to the relatively minor fluctuations in the system. Because periods of steady time are too brief for the drainage network or hillslope morphology to change, all variables may be regarded as independent except 10, the discharge of water and sediment from the system.

It will be seen that time can be considered as the most significant independent variable in landform studies, or regarded as of relatively little significance, depending upon the time-span involved. It is generally true to say that most modern geomorphological emphasis is upon studies concerned within graded or steady time.

Modern geomorphology

Like other branches of geography, the study and objectives of geomorphology have changed considerably since the end of the Second World War. Probably the greatest difference between today and the Davisian era is the emphasis placed upon morphometry, linked to the study of process. Geomorphologists generally are now less interested in establishing the denudation chronology of an area than in the careful measurement of the processes operating within it. Attention is paid not so much to the stage of the erosion cycle that has been reached as, for example, to the precise angle and composition of the slopes, measurement of the denudational processes acting upon them, and the discharge and load of the streams. In coastal studies the recognition of well-known coastal landforms is of little importance compared with the process involved in their formation. Questions concerning how (and how much) material is moved as longshore drift; movement in the nearshore zone by tidal streams and their residuals; waves and their behaviour, both in theory and through observation: these are the kind of problems which concern the modern researcher.

Interest in drainage basin morphometry has grown since R.E. Horton drew attention in 1945 to certain basic laws. The work of Horton has been built upon and extended since then and knowledge of the mathematical properties of drainage basins greatly extended. Today the case is widely argued for the recognition of the drainage basin as the basic geomorphic unit. This accords well with another trend, the study of applied geomorphology — applied, in the sense of the usefulness of the study to man. The most important single human resource is water, and the drainage basin is the most useful unit within which to study problems of supply, demand, transport, and water pollution.

One method is to consider the parts of an area as an interrelated working system, with inputs of energy, radiation and precipitation, and outputs of stream discharge and load. For this purpose the drainage basin is an ideal fundamental unit and the systems approach is now widely accepted as a useful way of organizing knowledge, because it provides a basis for the construction of geometrical and mathematical models of causally interrelated processes. The use of the drainage basin as a unit also permits flexibility for the purposes of study. It may be a first order basin consisting of a single short stream, or these may be combined into higher order basins of continually increasing complexity. A systems model might alternatively relate to the processes interacting upon a single slope; the various mechanisms of soil creep, precipitation and surface wash, removal of material in solution, deposition in the footslope area, all working together to produce a particular kind of profile. In contrast a section of coast might be considered where inputs of energy in the form of wind and waves move material from the offshore zone to the foreshore and then along the beach to create a series of constantly changing profiles. This coastal zone could be considered as a working system in which changes at one place, such as the construction of marine defences, react to affect beach conditions in other areas.

A further advantage in considering landforms as the product of a working system is that the attention is focused upon *all* the ascertainable factors that are associated in its formation, including man. Man has always been recognized as an instrument in the modification of landforms: the creation of reservoirs is an obvious example of how the flow, and therefore the erosion and transport capacity of a stream can be changed artificially. Sea defences are equally obvious in their effect in protecting the shoreline. The example (one of many) of the near destruction of Dawlish Warren at the mouth of the estuary of the River Exe highlights the importance of giving consideration to the whole system and not confining attention only to part of it. Less apparent, though equally important, are the indirect effects that changes in land use may have upon stream flow and load.

The processes involved in the modification of landforms are difficult to evaluate, partly due to problems of measurement and partly to the large number of variables involved when we try to examine them as a natural working system. It is often possible to measure with some degree of accuracy individual processes such as the solution and suspended load of streams, precipitation, and the rate of removal from exposed rock due to weathering. Other kinds of variable, like the amount of longshore drift or the movement of bed load in a stream, are much more difficult to estimate, although current research is now beginning to show results. When we try to study the natural interaction of all the variables together the problem becomes exceedingly complex, and may well contain unknown and possibly unsuspected factors to influence the result. For these reasons the laws of geomorphology are mostly expressed as hypotheses and are frequently statistical in nature. For example, Chorley and Kennedy (1971) believe that one measure of an equilibrium state may be the degree of statistically significant correlations which occur between variables. Fig. 1.4 is an example given by them with variables relating to slope geometry, basal stream, debris, and vegetation.

It has been seen above that normally only small areas can be regarded as homogeneous in the sense that they are in equilibrium. We may regard first order

Fig. 1.4 *Correlation of variables for slopes on the Charmontiers Limestone Plateau de Banigny, northern France.*
(a) *Shows where the actively working stream is adjacent to the bottom of the slope. The high degree of significant correlations appear to indicate that stream and slope processes are working in a degree of equilibrium*
(b) *Here the stream is not near the slope, correlations are few, and the element of equilibrium appears to be missing*

(a)

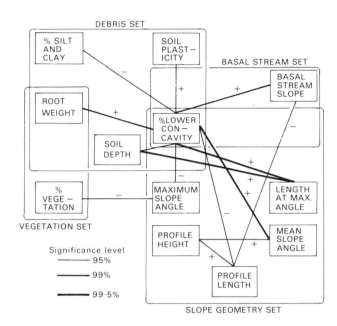

(b)

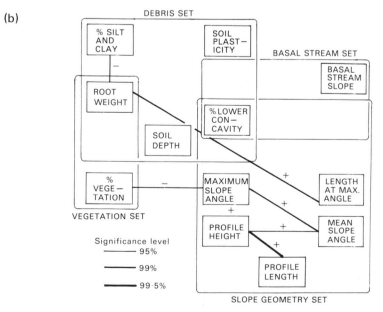

drainage basins as basic units (that may or may not be homogeneous, in this sense) which can be combined to form larger systems of higher order basins. Whatever its size, the drainage basin may also contain very small areas in a condition of steady state. Our approach may thus include on a macro-scale a consideration of a large drainage area as a working system with many other sub-systems as its component parts, or it may be that the system studied will be on a micro-scale and consist of one particular valley slope along a short reach of stream.

Until comparatively recently it had been widely accepted that the key to the past lay in the present: that if we understood the processes now acting upon the landscape, we should be able to use this knowledge to regress in time to reconstruct previous landscapes and thus determine the evolution of the landforms that we have today. There is now considerable doubt about this point of view. There are three reasons for the change of opinion:

1. Since we have begun to study process actively and quantitatively, the extreme complexity of landform evolution is becoming clear. So far we do not have the necessary knowledge about process to predict with any degree of certainty.

2. We are just emerging from a series of major glaciations, and therefore from a time of great change. Over the past several million years there have been great variations of process over much of the world, together with great local and eustatic fluctuations in sea level. These have resulted in considerable alterations in the base level of streams, whose capacity for erosion and transport has been changing from before the onset of the Pleistocene to the present. As a result of glacial and periglacial conditions, not only were normal processes interrupted but huge quantities of glacial, fluvio-glacial and aeolian material were deposited in various forms over the present temperate landscape. Much of the present load of our streams is composed of this material, which over periods of geologic time is an extreme rarity. The present characteristics of streams are therefore adjusted to load and sediment untypical of most of their history.

3. Man has increasingly affected the landscape over the last few thousand years, not only in constructing dams and weirs but in extensive deforestation, the building of large settlements, and the construction of drainage systems, all of which increase the rapidity of runoff. In so doing the natural relationships between precipitation and stream regime have been changed. Data obtained from the present behaviour, even of 'untouched' natural streams, cannot be used to determine their behaviour before human land use began.

The importance of the study of applied geomorphology has already been mentioned. This is particularly true in the area of environmental management. Recently concern has been growing all over the world for the quality of the environment in which we live. Any attempt at environmental management and control must necessarily be interdisciplinary. The geographer is especially well trained to undertake investigations of this nature, because geography is not concerned with the study of any particular categories of phenomena but with their related distributions in space. In this study the geomorphologist is playing an increasingly important role.

Consolidation

1. Fig. 1.2c on p. 7 shows the evolution of an area between two valleys. Do you recognize any of these profiles as existing in Britain, or typical of any part of the world? Plot these areas on a map. Can you suggest other physical similarities between them?
2. Discuss, with examples, the role of time in the study of landforms.
3. Why should the large number of significant correlations shown in Fig. 1.4 indicate that a state of equilibrium may exist?

Chapter 2

Drainage basin morphometry

The drainage basin as a unit of study

Geomorphology is concerned with the land surface of the whole earth. For the purpose of study it is essential that this surface is reduced in some rational way to a system of smaller units. Before 1945, in accordance with the general trend in geographic thought at that time, landscape was generally divided regionally into areas sharing the same principal geomorphic features. W.M. Davis recognized the unitary character of the drainage basin, but tended to distinguish stage of evolution as the basis of his field of study. Drainage basins differ enormously in size and vary from the largest of all, the Amazon, draining an area of some 3 810 000 km² with a complex network of tributary rivers, to a tiny catchment area measured in square metres and drained by a single stream. Thus an area enclosed by a major watershed may have within it a number of separate distinct catchment areas of varying size, but all contributing as part of the same main drainage system. The channels of the drainage network and the landforms they drain are bound together in a close causal relationship, in which any long-term change in the discharge characteristics of the streams will ultimately result in modification of the areas between them. Davis likened streams to the veins of a leaf, and the drainage basin to a whole leaf.

Today, for the following reasons, the drainage basin is generally regarded as the most satisfactory basic unit for study:

1. It is an areal unit defined by characteristics that can be measured quantitatively, thus providing an objective basis for analysis, comparison, and classification.
2. Drainage systems can be placed in an orderly hierarchy.
3. It can be treated as a working system with energy inputs of sunlight and precipitation, and outputs of stream discharge and load (see p. 68).

Difficulties can sometimes arise regarding the exact location of a watershed, or the definition of a stream (e.g. is it permanent, occasional, or possibly artificial?) depending on the nature of the terrain, and the scale of the map being used. In practice, arbitrary solutions to problems of this nature are generally satisfactory because any errors tend to be compensating. It is of great importance in the comparison of two basins that maps of the same scale and quality are used for each.

Fig. 2.1 *Diagrammatic representation of drainage basins. In* (a) *and* (b) *the arrangements of the drainage networks are different, as is the basin shape, but area and stream length are the same in each.* (a) *and* (b) *use the Strahler method of stream ordering.*

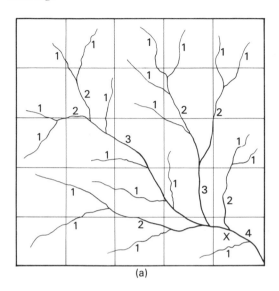

(a)

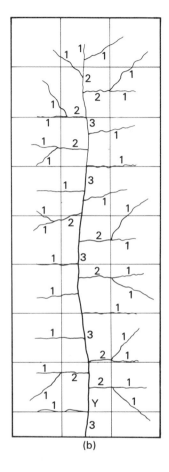

(b)

(c)

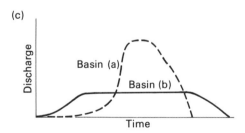

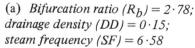

(a) *Bifurcation ratio* $(R_b) = 2 \cdot 78$;
drainage density $(DD) = 0 \cdot 15$;
steam frequency $(SF) = 6 \cdot 58$
(b) $R_b = 6 \cdot 50$; $DD = 0 \cdot 15$; $SF = 6 \cdot 58$
(c) *Potential flood discharge rate for* (a) *and* (b)
plotted against time (the flood hydrograph)
(d) *Horton's method of stream ordering.*
*Note that this requires the higher order
streams to be indentified upstream to their source*

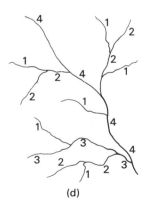

(d)

Fig, 2.2 (a) *Tributary streams in the five main basins of the upper Trent drainage area*
(b) *Stream numbers plotted against stream order on semi-log paper*

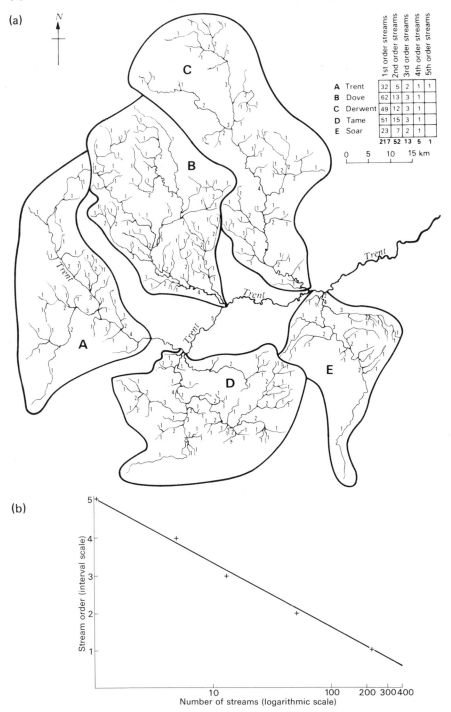

(a)

		1st order streams	2nd order streams	3rd order streams	4th order streams	5th order streams
A	Trent	32	5	2	1	1
B	Dove	62	13	3	1	
C	Derwent	49	12	3	1	
D	Tame	51	15	3	1	
E	Soar	23	7	2	1	
		217	**52**	**13**	**5**	**1**

0   5   10   15 km

(b)

Stream order (interval scale)

Number of streams (logarithmic scale)
10   100   200 300 400

Time-independent morphometry

Stream ordering

R.E. Horton proposed the allocation of order numbers to streams in 1945. His method is most clearly shown diagrammatically (Fig. 2.1d). Since then several different methods have been proposed. In many ways the simplest, and certainly the most widely accepted method today is that of Strahler, in which initial tributaries are allotted the rank of 1 and called first order streams. A second order stream is formed below the junction of two first order streams. A third order stream is formed below the junction of two second order streams, and so on (Figs. 2.1a and b). Note that the order number of a stream is unchanged irrespective of the number of tributaries of a lesser order which flow into it.

The drainage basin as a whole is normally allotted the number of the highest order stream that flows in it. For example, in Fig. 2.1, (a) is termed a fourth order basin, and (b) a third order basin.

Stream ordering reveals a rather curious attribute of natural drainage systems. If the number of streams is plotted against stream order on semi-log graph paper, provided the sample is sufficiently large, the plot will approximate to a straight line. Semi-log graph paper has an arithmetic scale along the horizontal axis, that is, a scale in which a given numerical difference is shown by a constant interval. The vertical scale however is logarithmic in which a given proportional difference is shown. If the proportion of the number of streams between one order and another remains constant the points plotted will lie in a straight line. A real situation is shown in Fig. 2.2a. The five drainage basins concerned exhibit a number of individual differences, yet when the drainage of the upper Trent is regarded as a whole it will be seen that the values recorded on Fig. 2.2b lie very close to a straight line. This attribute seems to hold good for most natural networks of permanent streams.

Recently many 'drainage' patterns have been randomly generated by computers. It has been found that if these randomly created 'streams' are ordered in the way described above and the numbers plotted on semi-log graph paper, the points will lie in a straight line. In other words in terms of the number of streams of each order, most natural stream networks very closely resemble one that is randomly generated.

It has been found that, if Horton's ordering system is used, the same kind of relationship exists between stream order and average stream length, and stream order and drainage area. Except that this time, length and area increase with order number. The establishment of these three relationships is variously called Hortonian analysis, Horton's laws, or the laws of morphometry.

The bifurcation ratio R_b

This is derived for any drainage basin by calculating the mean of the ratios between the total number of streams of each order and the total for the next order above it. For example, the bifurcation ratio for Fig. 2.1a, in which there are 20 streams of the first order, 6 of the second order, 2 of the third order, and 1 of the fourth order, is:

$$R_b = \frac{\frac{20}{6} + \frac{6}{2} + \frac{2}{1}}{3} = 2 \cdot 78$$

Similarly, it can be shown that the bifurcation ratio for Fig. 2.1b is 6·5. R_b is important because it is one of the factors which control the rate of discharge after sudden heavy rain. A little thought will reveal why this is so. Assume heavy rainfall of equal intensity and duration over the whole of basins (a) and (b) (different in shape but equal in area and stream length). The form of the drainage network in (a) is such that the tributaries will tend to concentrate discharge from most of the basin into the main stream in the area of X over a relatively short period of time. In (b) the drainage network is so arranged that discharge from tributaries is distributed relatively evenly over the course of the main stream. A substantial rise in water level at Y occurs sooner than in the case of (a), but peak discharge is reduced by being extended over a much longer time period. Fig. 2.1c shows the relationship between time and discharge rate for basins (a) and (b), and how the potential flood danger may increase as the value of the bifurcation ratio is reduced.

Drainage density DD and the constant of channel maintenance CCM

Drainage density is the total length L of all the streams in the basin divided by the area A of the whole basin, or $\frac{L}{A}$. It is thus the average length of stream channel for each unit area. The constant of channel maintenance is the reciprocal of drainage density, that is, it is the area of the basin divided by the sum of the lengths of all the permanent streams, or $\frac{A}{L}$. It shows the area required to maintain each unit length of stream. In Fig. 2.1, DD and CCM are the same for both (a) and (b), because stream length and area have also been made the same. They show that the

Fig. 2.3 (a) *A diagrammatic plan view of part of a drainage basin which has a CCM of 6 (km). That is, on average 6 km² of area support each 1 km of stream length. It will be seen that half the CCM (3km) represents the average distance of overland flow* (b) *Cross-sections through two drainage basins X and Y. Length of sections and slope angles are the same, but in Y the average length of overland flow is half that of X. t_1, t_2, and t_3 in X and Y are tributaries of the main stream. Surface runoff on Y will reach the streams twice as quickly as equal rainfall on X*

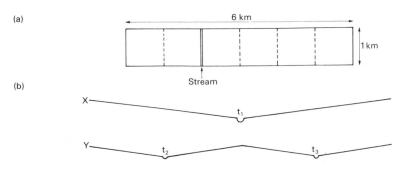

average length of stream for every unit of area is 0·15 and that 6·58 units of area are required to maintain one unit length of stream.

Because the *CCM* is the number of units of area supporting each unit of stream length, one half the *CCM* value gives the average horizontal distance between all the watersheds and streams within the basin (Fig. 2.3a). This is generally termed the length of overland flow *OF*. (A value very difficult to determine in any other way.) The speed of the unconcentrated flow overland is very much lower than when concentrated into a channel, probably at least five times slower. It follows that the smaller the value of *OF* the quicker surface runoff will enter the streams (Fig. 2.3b). In a relatively homogeneous area therefore less rainfall is required to contribute a significant volume of surface runoff to stream discharge when the value of *OF* is small than when it is large. It is worth noting here that in Britain, because rainstorms of the greatest intensity generally occur over a small area, small drainage basins are liable to more frequent and violent floods than large ones.

The hypsographic curve, and hypsometric curve and integral

The hypsographic curve
We have seen in Chapter 1 that the normal cycle of erosion concept is no longer generally considered wholly adequate. It is a **generic** model which provides both a generalized qualitative description of what the morphology of an area is like, and an indication of the stage of evolution it has reached. It is a **deductive** model in the sense that it is based on the assumed rapid uplift of a relatively level surface. There is obviously a need for another kind of model which will provide an objective description of landscape. We require an **empirical** model of the present landscape, based on quantitative evidence, and which is independent of time.

Fortunately such a model is available and is easy to construct. In 1940 Holmes used a hypsographic curve (Fig. 2.4) to show the proportions of the earth's solid

Fig. 2.4 *Hypsographic curve showing the areas of the earth's solid surface between successive levels*

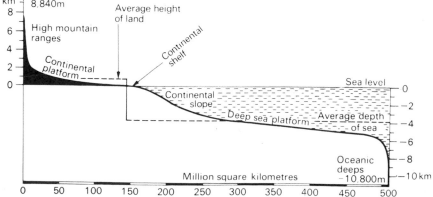

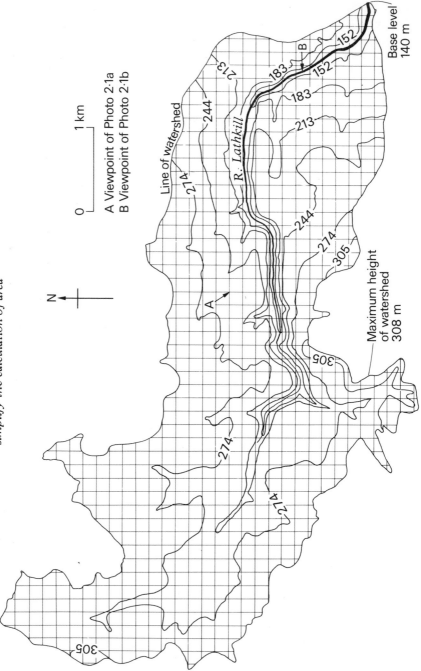

Fig. 2.5 Contour pattern of the drainage basin of the River Lathkill. The grid is to simplify the calculation of area

0   1 km

A Viewpoint of Photo 2·1a
B Viewpoint of Photo 2·1b

N

Line of watershed

R. Lathkill

Base level
140 m

Maximum height
of watershed
308 m

213

183

183

152

152

213

244

274

244

274

305

305

274

274

305

surface above or below different altitudes. This familiar technique shows that most of this surface is at two main levels, the continental, and the deep sea platforms. The curve is a summary, or model, of the whole surface of the earth. Similarly, it is also possible to draw a curve as a model of the surface of any given area of the earth, for example, a drainage basin. This is most conveniently done by using contour lines for the heights. The most accurate way to measure the area of an irregular shape is to use a polar planimeter. This instrument is expensive, however, and in practice it is normally quite adequate to superimpose a grid on the map (as in Fig. 2.5) so that the proportion of the area of the basin above a given contour may be estimated by counting squares.

Photos 2.1a and 2.1b show two aspects of Lathkill Dale in the Peak District. Compare the photographs with Fig. 2.5. It will be seen from the closeness of the contours beside the river how steep the valley sides are, and that as altitude increases, contours tend to be spaced wider apart, indicating the gentle slopes of the uplands.

The calculation of the hypsographic curve for a drainage basin requires first that we should find the area of the basin above (or below) a series of chosen heights, generally contour lines. The total number of squares in the Lathkill basin is estimated to be 540. It is necessary to say 'estimated' rather than 'counted', because the watershed boundary encloses awkwardly shaped parts of squares as well as whole squares. It is mostly a matter of judgement when adding up the total as to how many irregular parts are allotted to make one whole square. No two people will normally arrive at exactly the same figure, but differences should be very small and have little effect on the result. Precisely the same difficulty arises in estimating the number of whole squares above a contour line, but in practice estimation errors tend to cancel each other out. Fig. 2.5 exaggerates the counting difficulty because the scale has to be greatly reduced to fit the page of a book. (The scale of the map used in making the actual calculation was 1:25 000, and the grid squares were 5 x 5 mm.) It should be noted that it is not necessary to relate the size of the squares to the scale of the map, because we are working in proportions of parts of the area relative to the whole. The smaller the squares of the grid the more accurate will be the result. The choice of the size of grid is to some extent a compromise between speed and the degree of accuracy required.

The number of squares lying *below* the 152 m contour is estimated to be 10. Therefore the percentage of the basin *above* 152 m is

$$\frac{540 - 10}{540} \times 100 = 98$$

This figure is normally not given as a percentage, but in the form of a proportion between 0 and 1, in this case 0·98. Conventionally, the units of area (i.e. the number of squares) contained in the drainage basin are termed A, and the units of area *above* a given height are termed a. About 30 squares lie below the 183 m contour, therefore the proportion of the basin lying above 183 m is

$$\frac{a}{A} = \frac{540 - 30}{540} = 0·94$$

Photo 2.1 (a) *The relatively flat uplands of the Lathkill drainage basin. The river valley follows the line of trees in the middle distance running from right to left across the picture*

Photo 2.1 (b) *The Lathkill valley floor. Note the steep nature of the sides on the narrow valley. See Fig. 2.5 for the viewpoints of both photographs*

Proportions for the remaining contour heights are:

$$\text{Above 213 m} \qquad \frac{a}{A} = \frac{540 - 65}{540} = 0\cdot88$$

$$\text{Above 244 m} \qquad \frac{a}{A} = \frac{540 - 100}{540} = 0\cdot81$$

$$\text{Above 274 m} \qquad \frac{a}{A} = \frac{540 - 280}{540} = 0\cdot44$$

$$\text{Above 305 m} \qquad \frac{a}{A} = \frac{540 - 530}{540} = 0\cdot02$$

We now have a set of paired values: the contour height and the proportion of the basin lying above it. These may be plotted graphically to form a hypsographic curve. The curve for Lathkill Dale is given in Fig. 2.6.

By comparing the two axes of the graph in Fig. 2.6 it will be seen that 0·8 (i.e. 80 per cent) of the area of the drainage basin of the River Lathkill is high land above 244 m), and that the slope of the high land is relatively gentle and uniform. We also see that there is a very steep-sided valley, deeply incised into the uplands. (In this case there is only one main valley with very small tributaries (Fig. 2.5), but mostly there will be more than one — remember the curve is a generalization of the whole basin.) In other words, in the hypsographic curve we have a model of a drainage basin, arrived at by quantitative means, which describes the area objectively, and which is independent of time. It is therefore a statement of the present situation only.

Fig. 2.6 *Hypsographic curves for the drainage areas of the River Lathkill, Cocker Beck, and Car Dyke*

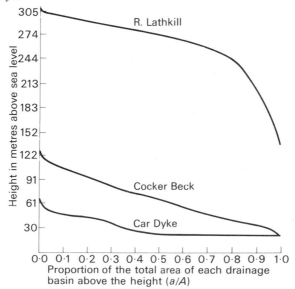

The hypsometric curve and integral
Before turning to a comparison of other areas there is one further useful calculation
which can be made using a similar technique. This is the derivation of the hypsometric integral, a numerical index of drainage basin morphology. To find the
integral we first construct a hypsometric curve by reducing the real heights of the
contours to a dimensionless ratio between 0 and 1 (in much the same way that we
converted areas above contours to the values a/A). To do so, find the difference
between the highest point on the watershed and local base level. In the case of the
Lathkill basin the difference is between 308 m and 140 m, i.e. 168 m. Then work
out the height of each contour above local base level and calculate this as a proportion of 168 m. For example, the 153 m contour is 13 m above local base level.
As a proportion of 168 m, on a scale between 0 and 1, it is $\frac{13}{168}$, which equals 0·08.
Conventionally, the difference between local base level and the highest point in the
basin is termed H, and that between any given contour and base level is termed h.
The proportion for the 183 m contour therefore is

$$\frac{h}{H} = \frac{43}{168} = 0·26$$

Similarly, proportions for the other contours are:

For the 213 m contour $\dfrac{h}{H} = \dfrac{73}{168} = 0·43$

For the 244 m contour $\dfrac{h}{H} = \dfrac{104}{168} = 0·62$

For the 274 m contour $\dfrac{h}{H} = \dfrac{134}{168} = 0·70$

For the 305 m contour $\dfrac{h}{H} = \dfrac{165}{168} = 0·98$

Once again we have a series of paired values, h/H and a/A. This time, however,
both run from 0 to 1, because both have been converted to proportions. Fig. 2.7
shows these values plotted graphically to form a hypsometric curve. The curve has
two main uses: for comparisons between basins (discussed below), and for the
derivation of the hypsometric integral. The integral is simply the area under the
curve expressed as a proportion of the area of the square of which the axes of the
graph form two sides.

The calculation of the hypsometric integral for the drainage basin of the River
Lathkill is given in Fig. 2.8, and comes to 0·71. The (hypothetical) assumptions are
(1) that at one time the whole area was a flat surface at the same altitude as the
present highest point in the basin, and (2) that the local base level remains constant.
The area of the square formed by the two axes of the graph above the curve thus
represents the amount of material so far removed by erosion. The area below the
curve, in this case about 71 per cent, represents the amount of material above local

Fig. 2.7 *Hypsometric curves for the drainage area of the River Lathkill, Cocker Beck, and Car Dyke. A curve approximating to the pecked diagonal would indicate about the same area of the basin at all heights. That is, a rolling landscape without distinctive flat areas or deeply incised valleys*

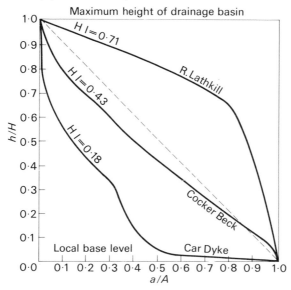

Fig. 2.8 *Calculation of the hypsometric integral for the area drained by the River Lathkill. Total number of small squares within the large square formed by the axes of the graph is 400, about 285 of which are below the curve. Therefore the hypsometric integral (HI) = $\frac{285}{400}$ = 0·71*

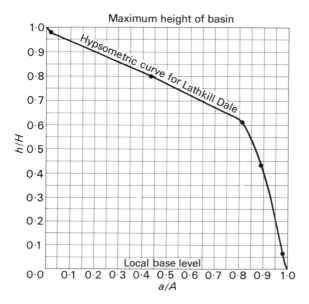

Fig. 2.9 *Location of the Cocker Beck and Car Dyke in relation to the River Trent*

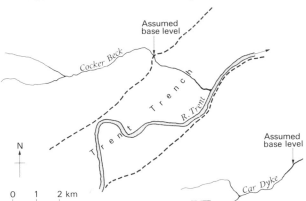

base level that still remains to be removed. The hypsometric integral is therefore a numerical description which may be used in the classification, and in the comparison of different landscapes.

The example of Lathkill Dale is taken from the Peak District, where one might expect to find the kind of landscape portrayed by the hypsographic curve described above, because it is carved out of massive Carboniferous Limestone. The morphology of much of Britain is less dramatic, so let us look at the application of those techniques to two areas of low relief.

Fig. 2.9 shows the location of two small drainage basins, those of the Car Dyke (Photos 2.2a and b) and Cocker Beck (Photos 2.3a and b), lying either side of the River Trent. The whole area covered by the map consists of Keuper Marl (a rock consisting of soft red marls interbedded with bands of sandstone called 'skerries'), dipping slightly towards the south-east. Both basins are roughly comparable in size, with the main direction of drainage from west to east, although the area of the Car Dyke system is a little larger than that of the Cocker Beck. Both have a similar assumed base level. In the case of the Cocker Beck, base level is assumed, for the purpose of this example, to be where the stream enters the Trent Trench. The Trench is Pleistocene in origin, and is the result of glacial meltwater and subsequent action by the River Trent. It would therefore be incorrect to include it in the characteristics of the drainage area of the Cocker Beck system.

Similarly, the lower Car Dyke which shortly joins the River Devon, enters an almost flat plain (also probably affected by glaciation) at a point where the watershed of the drainage basin becomes indistinguishable from those of other tributaries. Assumed base level has been chosen as the lowest point at which the Car Dyke drainage area can be determined with certainty.

Photo 2.3a gives an impression of the generally undulating landscape around the Cocker Beck, having fairly gentle slopes but little flat land. Photo 2.3b shows that this characterizes even the valley floor. The most notable feature is regularity. This general impression is confirmed by the map (Fig. 2.10). The contours are all about the same distance apart and more or less evenly distributed. The regularity of contour spacing, and the fact that they are not very close together, means the land is of a uniformly undulating nature with no very steep slopes. The long profile of

Photo 2.2 (a) *The upper part of the area drained by the Car Dyke system, looking across the stream towards the watershed*

Photo 2.2 (b) *The area close to assumed base level. The Car Dyke running from left to right across the centre of the picture is indicated by the line of bushes. The straightness of the line shows the flatness of this part of the basin. See Fig. 2.11 for viewpoints of both photographs*

Photo 2.3 (a) *View from the watershed across the drainage system of the Cocker Beck*

Photo 2.3 (b) *The valley floor. The location of the stream is masked by trees. See Fig. 2.10 for viewpoints of both photographs*

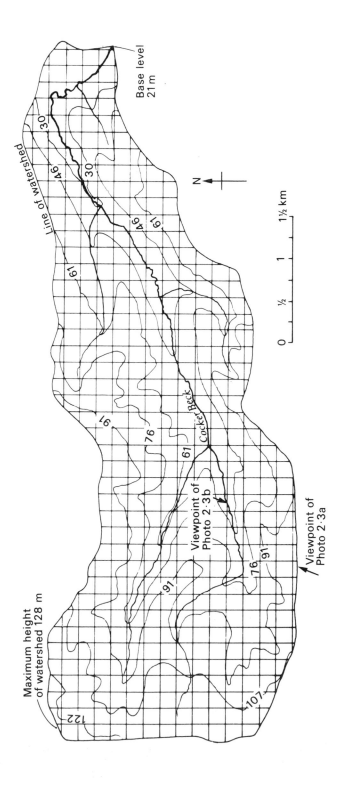

Fig. 2.10 *Contour pattern of the drainage basin of the Cocker Beck*

the valley descends some 107 m in just about 10 km, a drop of 10·7 m per km, sufficient to give the stream considerable energy after heavy rainfall. Looked at in Davisian terms, the area drained by the Cocker Beck seems to have most of the attributes of a landscape of mid-cycle maturity.

The differences in the hypsographic and hypsometric curves for Cocker Beck and Lathkill Dale (Figs. 2.6 and 2.7) are striking. It will be seen that the Cocker Beck approximates quite closely to the diagonal line in Fig. 2.7 showing graphically that about the same proportion of the drainage area is at any given height, and that there is no recognizable surface at any particular altitude. In the case of the River Lathkill there is the clearly prevailing upland surface, and the areally small but geomorphologically important lower area, consisting of very steep slopes. These differences result in a hypsometric integral for the Cocker Beck area of 0·43, compared with Lathkill Dale of 0·71.

Time-independent description

The hypsographic curve, hypsometric curve and integral are useful, quantitative, time-independent ways of describing the general morphology of a drainage basin. Interpretation however must be made with care. Both curves and integral average out the whole morphology of the basin to a single expression that may conceal considerable variation within the area. For example, a hypsometric curve approximating to the pecked diagonal in Fig. 2.7, with an integral value of about 0·5 could represent an undulating area of fairly complex drainage (the Cocker Beck basin is a simple example). Or it could represent a single valley with rectilinear valley sides sloping at a constant angle from watershed to base level. Indeed, it will be seen from Photo 2.1a and Fig. 2.5 that the uplands of the Lathkill area above the major break of slope where the valley sides rapidly steepen, do approximate to this sort of topography.

In Davisian terms the area drained by the River Lathkill is a landscape of youth. It has steep-sided V-shaped valleys, a high proportion of flat or gently undulating high land (probably an erosion surface) and very little flat low land. It may indeed *be* a youthful landscape; although first it is necessary to prove the erosion surface and date the uplift. The search for evidence of this kind may be interesting and rewarding, but cannot scientifically be linked to a term which describes the morphology of the area. What is required in all these cases is a time-independent method of landscape description of the kind suggested above.

The third sample selected for study, the area drained by the Car Dyke, is in marked contrast to the others (Fig. 2.11). It shares a common local base level with the Cocker Beck of 21 m O.D. The highest point is at 70 m O.D., but by far the greater part of the basin lies between 21 m and 46 m, a vertical difference of only 25 m. Photo 2.2a, looking towards the area of maximum relief, shows even here the general flatness of the landscape and the very gentle slope up to the watershed. Photo 2.2b is representative of 50 per cent of the drainage basin, in which the slope of the land is not detectable by eye, and only just sufficient to maintain the surface drainage. In appearance it is very much what we assume a Davisian peneplain in a landscape of old age would look like, and this is confirmed by the contour map and by the shape of the hypsographic curve.

We thus have two contrasting areas consisting of the same rock type, and sharing

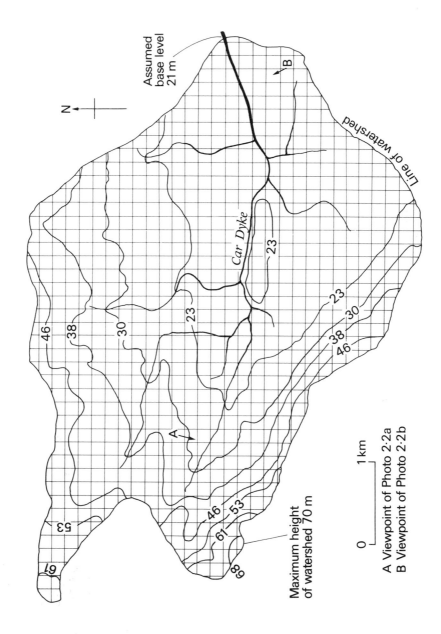

Fig. 2.11 *Contour pattern of the drainage basin of the Car Dyke*

similar base levels. Areas which, because they are less than 10 km apart can hardly have been affected by different processes of denudation. In Davisian descriptive terms the Cocker Beck is a mature (mid-cycle) type landscape, whilst the Car Dyke basin represents a landscape of old age. Yet the reason for the difference lies not in the length of time erosion has taken place, but in the history of the River Trent and the reduction in base level of the Cocker Beck. Before the Pleistocene the Trent ran well to the south of the Car Dyke. It can be established with some certainty (by

Fig. 2.12 *How the initial drainage pattern of the Cocker Beck system of streams was truncated during the Pleistocene by the excavation of the Trent Trench*

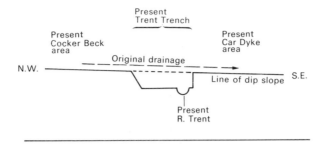

means of relicts of the ancient channel and fluvial deposits) that the slope of the long profile of the proto-Trent was gentle, and that it probably flowed through a landscape of low relief. The Trent Trench did not exist at this stage and the Cocker Beck seems to have flowed in a south-easterly direction to form part of the same drainage system as the Car Dyke (Fig. 2.12).

During the Pleistocene when the ice was at its maximum, the area was covered by an ice sheet. As the ice began to melt upstream, large quantities of meltwater were created which flowed down the old river valley. But in the area now occupied by the Car Dyke, and across the old river channel to the south, lay a mass of un-melted ice. The meltwater was forced by this ice to flow further north, and it excavated the Trent Trench as it did so, thereby considerably lowering effective base level by truncating the drainage pattern in the area of the present Cocker Beck. The present River Trent is left in this part of its course as a misfit stream in what was initially a meltwater channel.

We thus have two adjacent areas of the same age and rock type, and with similar local base levels, but which exhibit quite different morphological characteristics. This example (there are countless others) demonstrates the necessity for the use of descriptive techniques, objectively arrived at, and which are independent of time. This is an important consideration, since many modern geomorphologists think it probable that some areas may be regarded as remaining for long periods in a state of dynamic equilibrium — a concept considered below (p. 74). The methods presented here have been covered in considerable detail because they are particularly useful as alternatives to description in Davisian terms.

Consolidation

1. What are the advantages, and disadvantages, of using hypsographic curves, and hypsometric curves and integrals as descriptive models?
2. Some methods of drainage basin morphometry are described in this chapter. Can you suggest others that might be useful? For example, how could you compare the *shape* of two basins? (*Science in Geography 3: Data Description and Presentation*, Chapter 4, may help.)
3. Discuss the case for a time-independent method of landscape description. Exemplify your argument, using the techniques outlined above, applied to areas of which you have personal knowledge.

Chapter 3

Water: storage and flow

Storage, surface runoff, and groundwater

Surface storage

All the water from which streams are derived comes initially from atmospheric moisture, sometimes in the form of snow but generally as rain. When rain falls over a drainage basin a small proportion falls directly into the streams and marginally increases their discharge, but most of it falls on vegetation, or on bare earth or rock. A little of the rain may be intercepted by the vegetation and remain as droplets on leaves, and eventually evaporates to increase the water vapour content of the air. Most of the rain will reach the soil and, if this is not already saturated, soak in to form soil moisture. Part of this moisture is taken up by plants through their root system, eventually to be transpired back into the atmosphere as water vapour. Large quantities may be used in this way. Part of the soil moisture will evaporate directly from the surface of the ground. Because it is difficult to differentiate between water vapour derived as evaporation from soil and transpiration from plants, the total loss of water to the soil by this means is known as evapotranspiration. The amount of evapotranspiration taking place at any given time depends upon temperature, the relative humidity of the air, and the amount of foliage available — and thus mainly on the season of the year. Evapotranspiration will be at its maximum in a warm dry summer, and at a minimum in a wet and cold winter, when plants are not growing and deciduous trees are leafless. It is considerably increased by wind. It will also vary with the climatic pattern. Weyman (1975) states that in the United Kingdom in the extreme north-west Highlands, where the annual precipitation is 4000 mm, evapotranspiration is less than 350 mm, while in parts of East Anglia annual precipitation may be less than 500 mm and evapotranspiration up to 600 mm. In the driest areas of Britain therefore evapotranspiration may sometimes exceed precipitation.

The amount of water absorbed into the ground depends upon three factors: rock type, soil type, and land use. Unconsolidated gravels and sands (especially when these are sorted) have relatively large voids between the grains, and therefore soak up water most readily. (Even consolidated sandstone may absorb much water, for example, Bunter Sandstone can hold up to 28 per cent of its own volume.) Other than in well-jointed rocks, the ability to transmit water tends to decrease with particle size — when the clay fraction is reached the rock becomes impermeable. The soil itself also forms an important reservoir, depending again on the size of the pores between grains and the amount of organic material present. In a soil with a good crumb structure and plenty of humus, the surface layers may hold available to plants between 76 and 100 mm of rainfall equivalent.

Surface runoff

If rainfall is sufficiently heavy or prolonged the ground surface will become saturated when the amount of water reaching it is balanced by the rate at which it can be absorbed. Any rainfall in excess of this critical point will first fill up any surface depressions (which may provide an important capacity for temporary storage) and then result in surface runoff. Saturation does not necessarily take place evenly over the whole surface, especially on slopes. Here there is a tendency for the water to move downslope under gravity through pores in the soil and subsoil. This movement is above the level of the water table and is termed throughflow (page 39). The lower areas of a slope may thus become saturated and experience runoff before the upper parts. Land use is also significant in considering surface flow. The lack of vegetation (especially established woodland with its associated accumulation of leaf mould) is important. For example, when land has been cleared and ploughed, raindrops fall directly on to the soil, causing the surface particles to become compacted. This reduces the rate at which the water can be absorbed, thereby increasing (perhaps very considerably) the amount of runoff for a given amount of rain. Extensive areas of building, such as new housing estates, change the drainage characteristics of an area by providing roofs, roads, and an artificial drainage system to conduct any rainfall that takes place most quickly and efficiently into the local streams. The excessive flooding of the River Crouch around Wickford in Essex from 1960 has been attributed to the effect of the construction of Basildon New Town higher upstream.

Very rapid runoff during periods of heavy rain, often in Britain the result of summer thunderstorms, may result in large quantities of water being added to a stream very quickly. This may cause the stream to increase in depth rapidly and for a short time become a fast-moving torrent known as a flash flood. The Lynmouth disaster of 1952 was of this type. Another example of flash flooding is described by F.A. Barnes and H.R. Potter (*East Midland Geographer*, 1958). The flood occurred on the morning of 6 August 1957 in a number of small tributaries of the River Dove about 15 km west of Derby as a result of heavy thundery rain. Fig. 3.1 shows the drainage basin of one of these tributaries, the Foston Brook. The length of this stream from its confluence with the Dove (55 m O.D.) to its source (170 m O.D.), where it is known as the Bentley Brook, is about 15 km, draining a total area of about 32·5 km^2.

There was some rainfall on 5 August and the ground surface was already damp when heavy rain, estimated at between 125 mm and 200 mm fell in the area of Darley Moor (the centre of the storm) between 05.00 and 08.00 hours on 6 August. By 08.00 hours the Cubley Brook had risen between 1·83 m and 2·74 m above normal. The front of the flood, temporarily checked by an embankment, swept downstream towards Foston. It approached the bridge carrying the main A50(T) road from Derby to Stoke as a wave 1·8 m high. The bridge (Photo 3.1), consisting of three stone arches, of which only one was normally used by the Foston Brook, was completely submerged to a depth of 0·75 m and one arch collapsed within a

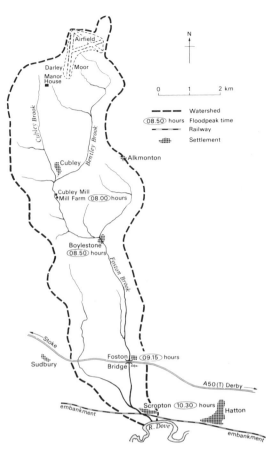

Fig. 3.1 *The area drained by the Foston Brook 15 km west of Derby*

Photo 3.1 *Foston Bridge after the flash flood. Note the strength of the stonework of which the bridge is made and the smallness of the Foston Brook in the foreground, still more than bank full, which did the damage*

Derby Evening Telegraph

few minutes and another shortly after. Barnes and Potter estimate the discharge at this time to have been over 170 m³/sec. Cottages upstream of the bridge were flooded to a depth of 2 m.

The destruction of the bridge at Foston so quickly and by such a small stream indicates the power of the flood. The bridge was an old one (exact dating is uncertain) and of stone construction. That the Foston Brook had been subject to flooding in the past is indicated by the presence of two arches in addition to that required for the normal flow of the stream. It is also obvious that prior to 1957 the bridge had been adequate to accommodate any floodwater. The rainfall on 6 August had been exceptionally heavy, but other factors are almost certainly responsible for the severity and speed of the flood. The whole drainage basin is on Keuper Marl (a clayey type of rock interbedded with sandstone); the higher area near the source of the Cubley and Bentley Brooks is formed by the Waterstones (a fine sandstone), while downstream is the Marl forming a relatively impermeable surface through which water could pass only slowly, and which therefore rapidly became saturated. Rain thereafter was conveyed to the streams as surface runoff. Before 1939 the relatively flat Darley Moor was badly drained and well wooded. During the war, advantage was taken of the level land to construct an airfield on the moor, involving extensive clearance of the woodland and a great improvement of the drainage into the upper tributaries of the Bentley Brook. There seems little doubt that this was a major factor in increasing the intensity and destructive nature of the flood.

Flash floods are important geomorphologically because the work the stream can accomplish in perhaps less than one hour may be greater than in a century or more of normal flow. The example of the Foston Brook has been given at some length to show the importance that land use may have, and especially the clearing of woodland or the improvement of land drainage (as in urban development), on the physical characteristics of a stream.

Groundwater

When rainfall is absorbed into permeable rock it continues to travel downwards and downslope under the influence of gravity until it meets an impermeable layer such as clay. A rock which can contain appreciable amounts of water that can be readily withdrawn is termed an aquifer. (Compare a rock like clay that also contains water sealed by surface tension around the minute particles of which it is comprised. This colloidal nature of clay makes it impermeable.) An aquifer may accumulate water until it becomes saturated. The upper surface of the saturated layer tends to follow below ground the approximate outline of the relief above, and is sometimes called the water table. If the water table reaches the surface, water may seep directly into a stream, or form a natural spring (Fig. 3.2a). The steepness of the water table depends upon the degree of permeability; it will tend to be steeper in less permeable rocks because these offer greater resistance to flow. The movement of water underground is rather complex and may be upward as well as downward as a result of hydrostatic pressure (i.e. pressure exerted by water at rest). Let us consider a lake of still water. Hydrostatic pressure at the surface will be zero and will increase regularly with depth. Because the surface of the lake is flat, pressure at any given depth (looked at another way, at any given altitude) below the lake surface, is the

Fig. 3.2 (a) *Storage and movement of water in an aquifer. The surface of the satu-rated layer is termed the water table. Throughflow is the term used for water moving downslope in and below the soil but above the water table*
(b) *Groundwater under pressure forming an artesian basin*

(a)

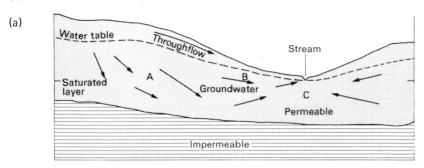

(b)

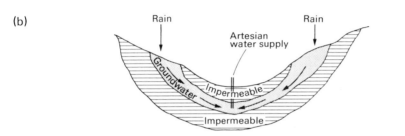

same. But the surface of the water table is not flat and hydrostatic pressure may therefore vary at points at the same altitude. In Fig. 3.2a, A is at the same altitude as B although pressure at A will be greater than at B. Obviously water in saturated ground will flow towards areas of lower pressure and so there will be a horizontal component of flow from A to B. Similarly groundwater at C, under pressure from water in places where the water table is higher, will tend to move upwards. If the aquifer is formed between two impermeable layers then hydrostatic pressure may be considerable and artesian water become available (Fig 3.2b).

Groundwater, flowing slowly through permeable material, accounts for perma-nent springs and represents a very large (and economically important) reserve to maintain streams in times of drought. The rate of flow of a stream maintained in dry periods entirely by groundwater is known as its *base flow*. As we have seen, in addition to natural springs water may enter streams directly (as in Fig. 3.2a) where the channel reaches the water table. In some cases a high proportion of the stream's natural supply may be provided in this way, and if the water table fluctuates in depth, generally between summer and winter, the stream may flow intermittently. Such streams exist in the chalk lands of England where they are often termed 'bournes'.

It is well known that water flowing initially along joint planes in massive lime-stone will, mainly by solution, enlarge small cracks eventually to form the classic limestone features of underground caves and river systems. It is now generally recognized that groundwater flowing through permeable material in or below the lower soil horizons will tend to exploit lines of weakness and gradually enlarge them to create 'pipes', down which the water will pass much more quickly. These may vary in size depending upon the stability of the material in which they are formed. The initiation of new surface streams may sometimes be the result of pipes which have become too large and have subsequently collapsed.

It is possible to approximate the speed of movement of water through soil or subsoil experimentally by timing movement through a sample in a container with waterproof sides, e.g. a plastic or metal pipe pressed into the ground, and the time taken for a known quantity of water poured into it to be absorbed recorded. Care must be taken that its natural compaction remains unaltered. The existence of natural pipes will greatly increase the speed of movement, and this factor must always be considered. Interesting experimental work can be carried out in a small stream's catchment area from measurements of rainfall, stream discharge (considered below), estimated evapotranspiration, and throughflow.

Transport by streams

Other than in areas of aridity or great cold, water is by far the most important agent of transport. Once canalized in a stream, moving water provides a rapid and efficient method of transport for the products of erosion. A surface stream normally flows at the lowest point of the cross-profile of its valley (there are exceptions), and thus forms the base level to which the slopes are graded. But a stream does not only transport, it may erode its own banks, deepen its own bed, or raise the valley floor with an accumulation of alluvial deposits. In so doing it changes the base level of the adjacent slopes, and hence in time the associated landform. The characteristics of streams and other processes which operate within them are thus of great importance to the geomorphologist.

Streams may be permanent, seasonal, or ephemeral. Permanent streams are those maintained by groundwater at all times. Seasonal streams are those which appear during wet seasons (e.g. the bournes referred to above) and are generally related to the height of the water table. Ephemeral streams are created by surface runoff in wet weather.

The **load** of a stream is the actual amount of material being transported. This may be carried in solution, by traction (the bed load), or in suspension. The potential load is the amount of material a stream can transport at any given point and changes with the volume and velocity of the water. This is in large part conditioned by rainfall. We must all have noticed after a dry spell the clear water of a permanent stream fed only from groundwater, compared with the coffee-coloured turbulence of the same stream after a period of heavy rain when surface runoff has greatly increased the depth, the velocity of flow, and the suspended load. Generally speaking the wider and deeper the water *within* the banks the greater the potential load a stream can carry. A stream is at its most efficient in terms of its carrying capacity when the channel is just filled by water to the top of its banks. This is known as a bank full condition. If the water rises over the banks and flows across the floodplain, friction will reduce the velocity (and therefore the energy) of the excess water and deposition will take place, the origin of natural levées.

The solution load

The solution load is most concentrated during periods of low rainfall. When a permanent stream is undiluted by surface runoff, it is maintained by groundwater which has passed slowly through soil and permeable rock, carrying with it in solution traces of organic and inorganic soluble material. In addition to naturally occurring substances, the fertilizers which farmers put on their fields may considerably augment a stream's solution load, especially in areas of intensive agriculture. So too will the disposal of industrial effluent and treated sewage. Although the environmental effects of this may be considerable, the removal of material in solution, whilst a very important denudation process, has no measurable direct effect upon the geomorphic characteristics of the river.

The bed load

The traction or bed load is the material which is moved along the bed of the stream by the tractive effect of the moving water. That considerable quantities of material are moved in this way there is little doubt, although the process is very difficult to measure in practice. (Some research is being currently conducted using concrete tanks dug into the bed of the stream, but results are not easy to interpret.) Sand, gravel, and boulder deposits in the beds of streams are proof that the process exists, and that under certain conditions it can be important. The minimum force required to move a particle of a given size is known as the **critical tractive force**. A number of equations have been suggested to calculate this, but for various reasons none of these is very satisfactory, and they are not considered here.

The **competence** of a stream depends upon the size of the largest particle it can move as traction load. It is comparatively simple to estimate a minimum value of competence by measuring the largest pebble or boulder that can be found in the stream bed that could only have been positioned as a result of stream action (i.e. a boulder which has not rolled down from the valley side). As we have seen, the volume and velocity of a stream can vary enormously with unusual meteorological conditions. Boulders of up to 10 m^3 were moved along the bed of the tiny West Lyn river during the North Devon flood of 15/16 August 1952, and a boulder weighing seven and a half tonnes was moved into the basement of a hotel. The Lynmouth disaster was the result of a freak storm. The flash flood that followed became a disaster because historic evidence, dating from the great flood of 1769 and visible in the form of boulders in the beds of the East and West Lyn demonstrating their potential competence in times of flood, were ignored. Building, from late Victorian times onward, was allowed to develop without perception of the flood danger. Bridges were made too small to take an exceptional flood and its associated load of debris. The archway of the bridge over the West Lyn is nearly four metres high, but this became completely blocked in 1952 and created an obstacle which diverted the stream and its load of boulders down the main street of Lynmouth. In the words of W.G. Green: 'The effect was like the work of a number of bulldozers'. Of British rivers only the Thames, with a catchment area nearly 100 times as large, has exceeded the estimated discharge of the River Lyn on that night, and then only twice since 1883. Yet there is some evidence to suggest that even the flood in 1952 was not as severe as the one of 1769.

The suspended load

The suspended load consists of fine material carried in suspension by the water itself. The amount of the suspended load depends partly on the quantity of material available to the stream and partly on the total volume of water involved. The size of the particles carried is determined by the velocity of the stream. Fig 3.3 shows the relationship between erosion, transportation, and deposition.

The curve of mean erosion velocity in Fig. 3.3 gives an approximation of the water speed required to move into suspension ('entrain' is the technical term) particles of diameter size varying from clay 0·001 mm to boulders of 100 mm. The mean fall velocity curve shows the water speed at which particles of a given size will fall out of suspension and be deposited. Both curves are prefixed by the word 'mean' because particles may be composed of varying materials having different densities. For example, heavier particles of a given diameter would require a rather higher velocity to entrain than those which are lighter. The curves thus represent the mean velocities for entrainment and deposition of uniform material appropriate to the products of erosion commonly found in rivers.

Two main points of interest arise from Fig. 3.3. Firstly, sand is moved at a lower stream velocity than either finer or coarser particles. The British Standard Code of Practice (1947), Fig. 3.4, defines sand as consisting of particles with diameters between 0·06 and 2·0 mm. The average erosion velocity curve shows that this is just the range over which the minimum velocity is required for entrainment, i.e. between about 12 and 15 cm/sec. Whereas clay particles with a diameter of 0·002 mm require the same high velocity of about 400 cm/sec to become entrained as large cobbles with a diameter of nearly 100 mm. If particles of a wide range of size are available to a stream normally flowing at a velocity below 15 cm/sec it follows that more sand will be transported than particles of any other size.

The second point which emerges from Fig. 3.3 is that the velocity required to maintain particles in suspension is less than that required to entrain them. For medium sand and finer material the fall velocity is very much less. (For the clay fraction it is near zero.) Thus once entrained the finer fraction of the suspended load can be maintained at very low stream velocities. It follows that the finer

Fig. 3.3 *The relationship between the velocity of stream water, and the erosion, transportation, and deposition of particles of different size (logarithmic scale)*

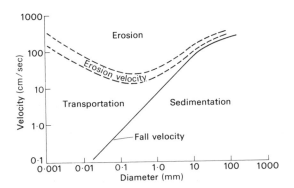

Fig. 3.4 *Classification of particle size in accordance with the British Standard Code of Practice (1947)*

Type	Size of dominant fraction (mm)	Field identification
Stones, boulders	200	
Cobbles	60–200	
Gravels	2–60	
Sand {coarse medium fine}	0·6–2 0·2–0·6 0·06–0·2	Particles visible to the naked eye. No cohesion when dry.
Silt	0·002–0·06	Particles mostly invisible to the naked eye. Some plasticity and dilatancy. Dry lumps possess cohesion but can be powdered easily in the fingers.
Clay	30 per cent of particle size less than 0·002	Smooth to the touch and plastic. No dilatancy. Sticks to the fingers. Shrinks on drying. Dry lumps can not be powdered but can be broken.

particles eroded and entrained by turbulent tributaries can be maintained in suspension by a less turbulent main stream.

It has been a commonly held fallacy that the speed of flow of a main stream must be less than that of its mountain tributaries. These may often be steeper, run over a more irregular bed, and consequently have a much greater element of turbulent flow. This is frequently apparent to the eye, but turbulence must not be mistaken for mean velocity. The eddies and splashing of a mountain stream give the appearance of rapid flow, and it is true that individual whirls in the water may indeed be very fast. The point to remember is that eddies do not contribute to the mean velocity of flow, because if motion is circular, although there is a component on one side of the circle flowing fast downstream, the component on the other side is flowing equally fast back upstream. The flow of the main stream with smoother bed and banks is generally less turbulent. The *apparent* speed of flow seems less, but the *mean* velocity and discharge is relatively high. When the discharge of the tributaries exceeds that of the main stream, the water first deepens and then floods.

The turbulence in tributaries can be important because the speed of moving water within an eddy or cascading over a rock can be sufficiently high to cause the entrainment of small particles. This is one factor which, after heavy rain, helps to account for the muddy appearance of main streams not sufficiently turbulent themselves for entrainment to take place.

It should be remembered that only a part of the suspended load may be material derived from a stream bed or banks. If rain is sufficiently heavy or prolonged for the soil to become saturated and direct runoff to take place, surface wash can entrain and provide transport for fine particles into streams where they enter directly into suspension. A very important factor leading to entrainment under these conditions is the 'explosive' effect of the impact of raindrops on bare soil. There are also other ways in which fine particles can become part of the suspended load of a

stream, for example, when a bank is undercut and portions of soil or subsoil fall
into the water. It follows that whereas the erosion velocity may account for the
entrainment of only part of the suspended load, deposition is wholly controlled by
the fall velocity.

For the mathematically minded, Stokes's law expresses the settling velocity for
small spherical grains:

$$V = \tfrac{2}{9} \times \frac{gr^3(d_1-d_2)}{u}$$

where V is the settling velocity, g is the acceleration of gravity, r is the radius of the
particle, d_1 the density of the particle, d_2 the density of the liquid, and u is the
viscosity of the fluid.

This formula should be used with caution because it holds good only for *small*
particles which are roughly spherical in shape. Calculations for large particles and
those of irregular shape become much more complex. In general the approximation
given in Fig. 3.3 is sufficient for most practical purposes.

The amount of material transported by rivers can be very great. The total
depends upon the size of the drainage area, the annual discharge of the system, and
the geological characteristics and structure through which the streams pass. Sometimes transport is affected by man, for example, in making dams and weirs, changing
the land use, improving river banks, and using fertilizers. Some figures for selected
rivers in the U.S.A. are given in Fig. 3.5a. The most significant value is probably
the total annual weight removed from each square kilometre of the drainage area.
Fig. 3.5b shows the proportions of suspended and solution loads of selected streams.
It should be noted that very great variations can occur under differing conditions.

An interesting comment is quoted by Cooke and Doornkamp (1974) on the
relationship between suspended load, climate, and man. Fig. 3.6 shows suspended
sediment yield plotted against mean annual runoff (a) for small drainage basins in

Fig. 3.5 (a). *Average annual suspended and dissolved (solution) load of four
American rivers*

River	Drainage area (km²)	Average discharge (m³/sec)	Average suspended load	Average dissolved load	Total tonnes/km² a year
			(million tonnes a year)		
Colorado near Cisco, Utah	62 400	786	13·6	3·99	282
Mississippi at Red River Landing, Louisiana	2 966 500	52 900	258·0	92·35	118
Delaware at Trenton, New Jersey	17 500	1 090	0·91	0·75	95
Bighorn at Kane, Wyoming	41 200	222	1·45	0·20	40

Source: Leopold, Wolman, and Miller (1964)

Fig. 3.5 (b) *The importance of the solution load in some areas*

| River | Percentage of load | |
	Suspended	Solution
San Juan River, Bluff, Utah	97	3
Pond Branch, Maryland	16	84
Volga, U.S.S.R.	36	64
Don, U.S.S.R.	45	55
Ob, U.S.S.R.	33	67
Wieprz, Poland	5	95
Derwent, Eddysbridge, U.K.	70	30
East Devon catchment areas	40	60

Source: Gregory and Walling (1973)

the U.S.A., and (b) for a sample of rivers elsewhere. There is a very remarkable peak in both curves for rivers in places where runoff is small, that is, in areas of semi-arid climate. The reason for this is that although the amount of erosion through runoff theoretically rises in proportion to volume, as runoff (and therefore precipitation) increase so does the vegetation cover, which becomes progressively more effective in protecting the soil. It is only when runoff becomes very great indeed that the suspended stream load begins to increase once more. Indeed it will be seen from curve (b) that suspended sediment under very humid conditions does not equal that in semi-arid areas until the amount of runoff has increased by more than 2000 per cent.

Equally important is the height of the peak for curve (a) of the rivers in the U.S.A. It is suggested that the reason for this is because human activity has more greatly affected the vulnerable environment of the semi-arid lands there than in other countries – a good example of the way in which the works of man may greatly affect physical processes.

Fig. 3.6 *Relationship between suspended sediment yield from streams and runoff* (a) *in the U.S.A., and* (b) *for a sample of other rivers*

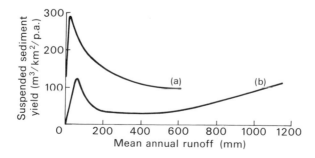

Stream flow and channel shape

Theoretical considerations

When water flows down a straight stream channel it is operated upon by two forces, that of gravity, and friction with the bed and sides. Gravity (g) is a force which acts as though exerted from the centre of the earth. On a plane that is not horizontal, i.e. on a slope, the downslope component of the gravitational force varies in proportion to the sine of the angle the slope makes with the horizontal (Fig. 3.7). The amount of friction present depends upon the roughness of the channel sides and bed. Roughness is determined by the height to which protuberances or 'roughness elements' stick out from the bed and sides of a theoretically smooth channel. These may vary from fine particles of silt or clay, through sand and gravel to large boulders. Vegetation also increases friction. The higher the roughness elements the greater will be the friction and therefore the drag on the water in contact with them, and the greater the energy required to overcome it. Resistance to the downslope flow of water is the ratio between stream depth D, and the height of the roughness elements k. The larger the value of D/k the smaller will be the resistance.

Fig. 3.7 *The effective gravitational force F down a slope at angle θ to the horizontal*

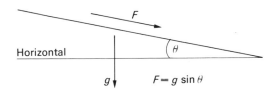

Fig. 3.8 *Measurements of the transverse cross-section of a stream. w is the width of the channel at the water surface XZ, p is the wetted perimeter, i.e. the length of the cross-section profile actually in contact with the water. D is the depth of the water. Bank full is when the water surface rises to $X_1 Z_1$. The hydraulic radius R is the ratio between the area of the cross-section of the water and the wetted perimeter, or a/p. Note that when channels are wide and the water is shallow, D and R are almost equal.*

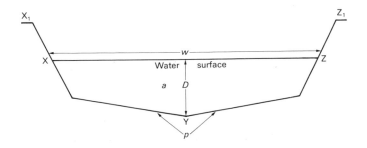

Fig. 3.9 *Manning's roughness coefficients*

Wetted perimeter	Coefficent
Earth canal, straight, good condition	0·020
Streams, earth canals, fair condition, some moss growth	0·025
Winding natural streams, earth canals in poor condition, considerable moss growth	0·035
Mountain streams, rocky beds; winding streams with variable sections; vegetation on banks	0·040–0·050

The theoretical mean velocity of flow in a natural stream is generally still calculated from Manning's formula of 1889:

$$v = 1 \cdot 5 \frac{R^{2/3} S^{1/2}}{n}$$

where R is the hydraulic radius a/p (Fig. 3.8), S is the sine of the angle of slope, and n is a roughness coefficient, determined empirically by Manning, some values of which are given in Fig. 3.9. The choice of which coefficient to use for a given part of a stream is a matter of judgement. Nevertheless, it has proved a useful approximation and is still widely used.

We have seen that friction with channel bed and sides slows down the layer of water in contact with them. This layer may even be stationary in places. Water in the centre of a straight channel will normally have the greatest velocity. The effect of drag caused by friction is transferred with decreasing effect from the contact layer to the faster flow at the centre through the viscosity of the water. Fig. 3.10 shows two hypothetical transverse cross-sections of straight river channels and the typical velocity distribution which might be expected.

Fig. 3.10 *Typical velocity distributions in two hypothetical stream channels. Note the effect of the narrow deep channel* (a), *and the more commonly occurring shallow channel* (b). *Velocities are in cm/sec*

(a)

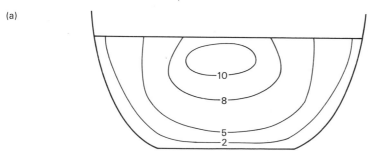

(b)

To obtain the discharge of a stream it is necessary to find the mean velocity of the water and to multiply this by the area of the cross-section. In practice it has been found that the mean of two readings taken in the centre or deepest part of the channel, one at 20 per cent and the other at 80 per cent of total depth give a close approximation of the average velocity in the cross-section. Rates of discharge are normally given in cubic feet per second (cusecs) or in cubic metres per second (m³/sec).

Practical considerations
The close observation and study of streams is an interesting aspect of geomorphology, and of immense importance in the study of processes affecting landforms. The practical difficulties of instrumentation are not as formidable as might seem at first, provided the stream selected is not too large. Much information on the behaviour and characteristics of the flow of water can be obtained from a stream about 0·5 m in depth and 2 m or 3 m wide.

The profile of the channel cross-section is easily obtained (preferably after a dry spell when the water is low) by using a measuring pole, and a wire or cord tightly stretched from one bank to the other. The angular measurement of the long profile may be more difficult over short distances, especially if the stream bed is uneven. In practice the slope of the bed is assumed to be that of the surface of the water flowing over it. The fall over a given distance can be measured by driving two measuring poles of equal height into the stream bed at A and B (Fig. 3.11a) so that the water level is the same height on both poles. Place a level horizontally on pole B and read back the height of the level on pole A. The difference between this observed height and the top of pole A is the amount of fall in the long profile between A and B.

Fig. 3.11 *Measurement of the gradient of a section of the long profile of a stream. For explanation see text*

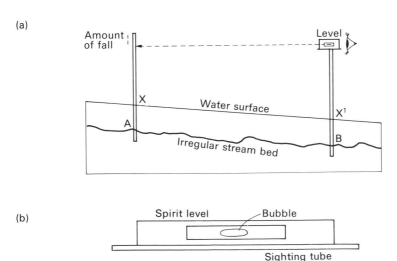

A surprisingly accurate sighting level (Fig 3.11b) can be readily constructed from an ordinary spirit level and a piece of tube if a standard instrument is not available. The case of the spirit level should be rectangular in shape. The tube may be made of any rigid material, including glass, with an internal diameter of about 5 mm or slightly less. A convenient length is about 20 cm. The tube is then fastened securely to the side of the level, exactly parallel to the glass tube containing the spirit bubble. The tube is then used like the sights of a rifle. Three people are necessary for this method: two at pole B, one to ensure that the level is horizontal by keeping the bubble central, and one to take the sight. It is desirable to have someone at pole A to place a finger on the precise height observed from B. (Because at any distance figures may be difficult to read through the sighting tube, whereas a finger is easily seen.) This technique can readily be adapted for the measurement of slope and beach profiles, or for contouring and recording spot heights over relatively small areas.

Water velocity is normally measured using a current meter in which the speed of a rotating propeller or set of cups is converted into metres per second. A sophisticated (and expensive) meter of this type may not be available, or its use impractical. Fortunately, it is again possible to improvise. If it is desired to estimate the discharge of a stream, it is first necessary to calculate the mean velocity at a given place. It is known that the maximum velocity in a straight channel will be found near the centre of the deepest part of the channel, at or just below the surface. To estimate the maximum velocity, take a stick and tie a weight (a stone or bit of metal will do) so that it will float upright in the water like a fisherman's float. Take two points a known distance apart along the stream; drop the float as nearly as possible into the deepest part of the channel, and time it over a known distance. This will provide the maximum velocity of the water. To find the mean velocity of the whole stream

Fig. 3.12 (a) *A simple device for measuring water speed at the surface. If the nails are allowed to touch the water, when the surface velocity reaches 0·22 m/sec ripples will form. The length of l will vary in proportion to the surface velocity* (b) *The velocity in m/sec for a given length of l*

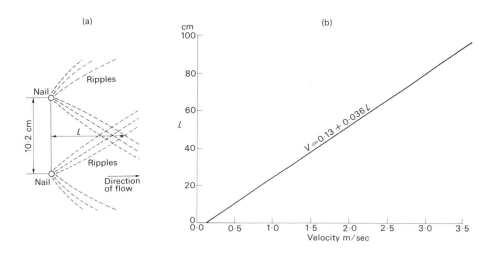

it is generally accepted that a good approximation may be obtained by multiplying the maximum velocity by 0·8.

A very simple device for measuring maximum surface velocity of a fast flowing stream is described by Mr. E.C. Thrupp in the *Minutes of the Meetings of the Institute of Civil Engineers* (1907), p. 216. It consists of a piece of wood with two nails driven through it exactly 10·2 cm apart. If the nails touch the surface of a stream it is known that when the velocity of the water reaches 0·22 m/sec ripples will form. If the piece of wood is held above the surface perpendicular to the flow, the ripples from each nail will converge downstream (Fig. 3.12a), and then diverge. The distance between a straight line joining the nails and the point of divergence will increase in proportion to the speed of flow. The relationship is given by the equation:

$$v = 0·13 + 0·036l$$

where v is the velocity of the water in m/sec, and l the distance in cm from the line of the nails to the point of divergence of the ripples. Water velocity for a given value of l can be quickly determined using the graph at Fig. 3.12b.

The study of the distribution of water velocity over the cross-section is important, especially when the channel is irregular either in cross-section or long profile, because this affects the erosion characteristics of the stream. A number of measurements for each cross-section are required. For these some sort of submersible current meter is essential. If one of the standard types of meter is not obtainable a substitute is easy and cheap to construct from readily available materials. Fig. 3.13 shows such a home-made meter. The U-tube, made of glass, is partly filled with coloured paraffin. When both ends of the tube are placed in moving water parallel to the direction of flow the air is subject to compression at one end and suction at the other, thus raising the height of the paraffin in one side of the U-tube relative to that in the other. This difference in height is a measure of the velocity of the water.

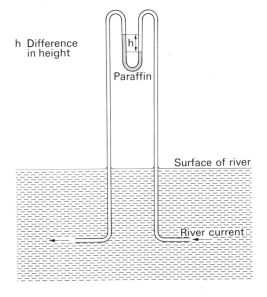

h Difference
in height

h

Paraffin

Surface of river

River current

Fig. 3.13 *A home-made current meter*

The meter has to be calibrated, i.e. the differences in height have to be converted to metres per second. (The method for doing this is given in Hammond, R. and McCullagh, P.S., *Quantitative Techniques in Geography* (O.U.P., 1974), pp. 233-6.) Although theoretically not quite as accurate as a standard instrument, the meter works reasonably well in practice.

Turbulent flow

The flow of any liquid may be of two kinds: laminar or turbulent. Laminar flow is when layers of water move over each other at different speeds, rather like sliding plates. Laminar flow is common in groundwater, but never occurs in natural streams, except possibly in a thin film in contact with the bed and banks. Therefore laminar flow is not considered here.

Turbulence consists of erratic eddies within the main flow. Water actually in contact with the wetted perimeter, and thus most affected by friction, may be stationary or nearly so, with velocity increasing away from the perimeter as the effect of the friction decreases towards the centre of the deep channel. It has been calculated that between 95 and 97 per cent of the kinetic energy of a natural stream is used in overcoming friction, not only with the perimeter but also between turbulent eddies in the water itself. (It is this element of friction within a liquid which determines its viscosity.) When the mean velocity is relatively low, turbulence is reduced and may not be apparent to the eye, except perhaps close to the edge. As the water level rises the mean velocity increases, because the *area* of the cross-section increases in proportion more than the length of the wetted perimeter, and the friction effect is reduced. This is why the quiet flow of a river when the water level is low can be transformed into a rapidly moving torrent when the water surface rises to near bank full conditions. The degree of turbulence is a product of channel roughness and water speed. As velocity increases erratic turbulent eddies may be seen on the surface. Turbulence is the mechanism that maintains particles in suspension in water. The faster the flow the greater the turbulence, and the larger the particles that can be carried. As we have seen the velocity of turbulent eddies may play an important part in the initial entrainment of suspended particles.

Turbulent flow is of two kinds: a slower rate of flow termed tranquil (or lower regime), and a faster rate of flow termed fast (or shooting, or upper regime). Tranquil flow consists of the ordinary kind of turbulent flow found in all natural streams, however slowly the water is moving. But as a critical velocity is passed, for example if the gradient of the bed steepens, or where the width of the channel is reduced, the flow changes its character and becomes fast. This change is related to the natural speed of a low wave in shallow still water, and was first demonstrated by a horse on the Glasgow to Ardrossan canal in Scotland in 1834. The horse, harnessed to a barge, took fright and bolted along the towpath taking the barge with it. The movement caused a wave to form in the canal and it so happened that the natural speed of the wave and that of the horse exactly coincided. The barge was thus carried along on top of the wave, greatly reducing friction with the water. The result was that the horse was able to pull the barge very much further than would have been physically possible if it had been running slightly faster or slower. Subsequent investigation showed that the speed of a low wave in still water (as in a canal) is determined solely by the depth of water, and can be calculated using the formula

$$c = \sqrt{gD}$$

where c is the velocity of the wave in m/sec, g is the acceleration due to gravity (9·81 m/sec), and D is the depth of water in the canal in metres.

The critical point at which flow changes from tranquil to fast is when the average velocity (v) of stream in an open channel reaches the same speed as a natural low wave would travel in still water of similar depth. That is, when $v = \sqrt{gD}$. For example, in a stream 1 m deep the critical point is reached when the average velocity equals $\sqrt{9·81 \times 1}$, that is when $v = 3·132$ m/sec. This relationship is known as the Froude, or F, number. But because the depth of water in a natural stream varies over the cross-section, the hydraulic radius R (Fig. 3.8) is generally substituted for D. The value of F may thus be determined from the formula

$$F = \frac{v}{\sqrt{gR}}$$

The critical value of F is 1. With $F < 1$ the flow is tranquil; with $F > 1$ the flow is fast. For example, in a stream with a hydraulic radius of 2 m flowing at a mean velocity of 1m/sec

$$F = \frac{1}{\sqrt{9·81 \times 2}} = 0·23$$

and the flow is tranquil. For this to change to fast flow the mean velocity of the water would have to reach nearly 5 m/sec.

As the critical value of $F = 1$ is exceeded, the *velocity of the water accelerates rapidly* and the erosional power of the stream is greatly increased. At the same time because of the increase in speed, the depth of water is reduced. (This characteristic change in flow is termed 'the hydraulic jump'. Because the velocity of the water exceeds the speed of a natural wave it is no longer possible for waves to move upstream. If a stone is cast into a stream in this condition the waves formed will all move downstream. Like an aeroplane overtaking the speed of its own sound waves as it passes through the sound barrier, at $F > 1$ the water overtakes the speed of any wave form within it.) If the gradient of the long profile diminishes, as at the lower end of a fast reach, or the width of the channel increases, the critical value of $F = 1$ may again be passed. As flow returns to tranquil the velocity of the stream is reduced, causing the depth to increase abruptly. This sudden elevation of the water surface often takes the form of a standing wave, sometimes with a reverse flow of water within it, and forms the dangerous stop wave well known to canoeists.

The ability of a stream to cause erosion depends upon its kinetic energy, that is, the energy of the moving water. The faster the flow the greater the kinetic energy. As the point of $F = 1$ is exceeded more energy is concentrated on a shorter wetted perimeter, thus greatly increasing erosional potential. Because of the increased erosional power of a stream under fast flow conditions, the places where this is likely to occur most often will be where most energy will be available to move bed load and to entrain small particles. When the width of the channel of a tranquil flowing stream is reduced, for example when it flows through a gorge cut into hard rock, in order to accommodate the volume of water in the reduced cross-section the speed increases. Under these circumstances the critical $F = 1$ value may be passed and flow become fast. The size of boulders found in gorges of some relatively small streams bear witness to the energy developed under fast flow conditions.

So far we have only considered the erosional properties of permanent streams. Surface runoff can also be very important as a denudational process in the entrainment of small particles. Runoff may occur as a sheet of water moving downslope, but this condition is comparatively rare. Generally surface water will form rivulets (ephemeral streams) until rainfall ceases. These streamlets form little gulleys on the surface, and are often an important way in which soil is removed. Vegetation is an important factor in inhibiting gulleying which is generally most pronounced on a bare unconsolidated surface. Considerable energy can be developed by small streams of this nature. This is because they are shallow, and therefore the critical value $F = 1$ is exceeded at low water velocities. For example, a small surface stream with a hydraulic radius (R) of 5 mm (0·005 m) flows at a rate of 250 mm (0·25 m) per second. Calculate the F value:

$$F = \frac{v}{\sqrt{gR}} = \frac{0·25}{\sqrt{9·81 \times 0·005}} = \frac{0·25}{0·2215} = 1·13$$

Thus a typical ephemeral stream conveying surface runoff about 6 mm deep with a velocity of only 25 cm/sec has already passed the critical F value and is in a state of fast flow.

Erosion, deposition, and negative feedback

During dry periods when streams are supplied only from groundwater springs, the depth of water is reduced, erosion is at a minimum and for all practical purposes can generally be regarded as zero. Transport of the bed load stops. No particles are entrained to form a suspended load, and unless the stream receives material already in suspension from a tributary, the water will be clear and the flow tranquil. (This applies relatively even to mountain streams with steep gradients.) The only work done by a stream in this regime is the transport of its solution load.

As depth increases after periods of rainfall water velocity and turbulence also increase, together with the length of the wetted perimeter. Energy becomes available for transport and erosion. At a critical velocity material on the bed begins to move and banks may be attacked to add to the load. As we nave seen (page 40) the efficiency of the stream as an agent of erosion and transport increases until bank full condition is reached. Stream energy is not reduced either by solution or suspension loads. In fact one rather unexpected result of water's acquiring suspended material is that it actually increases the stream's energy by reducing turbulence and therefore internal friction. It has been shown experimentally that, over the same bed form, friction can be reduced by 28 per cent through the addition of material in suspension.

Fig. 3.14 *Water velocity and the movement of dunes and anti-dunes in a stream with a sandy bed*

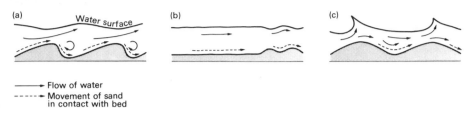

——————▶ Flow of water
- - - - -▶ Movement of sand
          in contact with bed

A feature of streams having a sandy bed is the formation of dunes (ripples) and anti-dunes. A natural stream rarely has a flat bed. If sand is transported along the bed, dunes are formed by moving water in much the same way as by wind on land (see p. 105) except that critical velocities are much smaller, and water may carry some sand in suspension. As the sand is pushed over the top of the dune and down the steep (slip) face, the dunes can be seen to migrate slowly downstream. At low velocities, with a correspondingly low value for F, grains of sand move along the bed individually. As the value of F increases the dunes grow in size, offering greater resistance to the water. Turbulence grows and may begin to show on the surface (Fig. 3.14a). As the velocity of the stream increases, a critical stage is reached when the dunes are washed away and whole groups of sand grains move downstream together (Fig. 3.14b). The removal of the dunes results in a considerable lowering of resistance to the flow of the water with an increase in velocity and the carrying capacity of the stream.

If the speed of the moving water increases still further, anti-dunes may form (Fig. 3.14c) while the surface of the water will have a series of standing waves tending to break upstream. Movement of sand is still downstream, but this time grains are stripped rapidly from the downstream face of one dune and piled onto the upstream face of the dune below. Anti-dunes are so called because this kind of sand movement causes the dunes to migrate upstream, although movement of the sand is downstream.

Under experimental conditions with a depth of water of between 0·12 m and 0·30 m dunes have been produced with heights ranging from 0·05 m to 0·30 m. In a very deep river, dune heights of over 9 m have been recorded.

Discharge in any stream changes continually throughout the year, and may vary considerably within hours or even minutes. As the quantity and velocity of water in the channel changes so the erosional and depositional properties of the stream change too. Any given reach will sometimes be subject to erosion and sometimes to deposition, and sometimes to periods when neither attribute is present to a measurable extent. A stream is said to be in a state of equilibrium (or at grade) when the energy available for erosion and transport (i.e. energy in addition to that required for overcoming friction with the wetted perimeter and within the water itself) is exactly balanced by the amount of material received.

The characteristic channel form of any stream is related to the average discharge over a period of time. For example, violent storms, resulting in a period of stream bed erosion, are followed by a reduced flow in which deposition tends to restore the balance. Equilibrium will thus be maintained provided that the average amount of water available to the stream, and therefore the average discharge, remains unaltered. It has been seen that streams are at their most efficient (that is, possess the highest energy for every unit of discharge) when they are in a bank full state. It is important, therefore, if equilibrium is to be maintained, that the average number of times a bank full condition is reached does not materially alter over a span of years. That such differences do occur is well known. One obvious reason in the long term could be climatic change, resulting in alterations in the amount of mean annual rainfall. There are other possible causes, among them the effect man's activities have upon land use (e.g. deforestation, land drainage, and building).

The effect of a long-term decrease in discharge is to reduce the energy of the stream and diminish its capacity to transport bed load, or carry large particles in suspension. Deposition will therefore take place. The effect of an increase of stream energy available for erosion and transport is not so easy to predict. It had been assumed by many in the past that an increase in discharge would invariably result in bed erosion and a deepening of the channel. It is now known that this does not necessarily happen, because other variables are involved. Take an obvious example in which a stream has a bed of resistant rock with banks consisting of mixed gravel and alluvium. The stream will adjust by eroding its banks until it is possible to accommodate the increased volume of water. But even in the case where the channel is in relatively homogeneous material the interrelationships are so complex that it is not yet possible to predict whether the stream will react by an increase in depth, or width, or both.

The reason why a natural stream will tend towards equilibrium with its surroundings after a sustained increase in discharge is due to negative feedback relationships among the variables concerned. Feedback, in the sense used here, is the ability of a stream, or drainage system, to regulate itself automatically to changed conditions. Feedback in fluvial relationships is negative. That is, as discharge and erosion increase the opposing forces resisting the flow of water increase too until a new equilibrium is achieved by the enlarged stream.

The mechanism of negative feedback works like this. When discharge increases, the wetted perimeter is enlarged by the erosion of bed, or banks, or both. As the wetted perimeter lengthens so friction increases proportionately, absorbing the energy available for erosion until a point is reached when erosion is reduced to a minimum and equilibrium again restored. In addition to the fixed perimeter much energy may also be used in transporting the bed load, especially if this is composed of gravel, or larger stones or boulders. Under these conditions so much energy may be absorbed in transport that little enlargement of the area of channel cross-section may take place. (The presence of a bed load with a high energy demand may thus lead to an increasing tendency to flood.) Similarly in a stream with a sandy bed the formation of dunes (but not anti-dunes) will add greatly to perimeter friction. The importance of negative feedback in fluvial processes and the role it plays in the concept of dynamic equilibrium is demonstrated when a valley is considered as a whole working system (p. 68).

Consolidation

1. (a) Why does the level of the water table tend to follow surface relief?
   (b) Explain how water moves upwards as well as downwards in an aquifer.
2. What natural circumstances may cause a stream not to flow at the lowest point of its valley cross-profile?
3. Describe ways in which a stream can acquire a suspended load. How does turbulence help this process?
4. Why may the presence of a bed load with a high energy demand be a factor which tends to increase the frequency of floods? Why may the bed load otherwise be important?
5. What is meant by 'fast' flow? Fig. 3.15 shows a hypothetical river channel. What must the speed of the water be for flow to change from tranquil to rapid?
6. What do you understand by a stream's efficiency? When is it at maximum? And why?
7. Waterborne deposits are always sorted. Using Fig. 3.3 describe and explain this process in detail.
8. Fig. 3.5 shows that the average proportion of suspended and solution load varies considerably between different rivers. Suggest reasons why this should be so and indicate how they apply to the rivers shown.

Fig. 3.15 *Dimensions of a hypothetical stream*

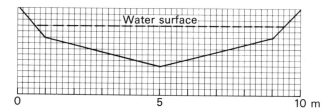

Chapter 4

Slopes

Weathering

It is only comparatively recently that the importance of weathering in the modification of the landscape has been fully realized. Today a great deal of research is being undertaken in this new field. All landforms that we see around us are to a greater or lesser extent the result of weathering. Before the rock that forms the backbone of the hills can move under the influence of gravity to create the slopes and surfaces we see in the landscape, it first must be broken into pieces. It has been customary historically to divide this process rather arbitrarily in two: mechanical weathering and chemical weathering. Today both are recognized as being so closely connected that it is generally not possible to consider one without the other. Even the most obvious kind of mechanical weathering, that of freeze/thaw action which produces the angular fragments of which some scree slopes are formed, is assisted by chemical action. Initial weaknesses of many kinds permit water to penetrate the rock and therefore chemical action to take place. The altered products may themselves expand and in so doing cause fractures, or they may create or enlarge cracks and spaces where more water can collect and cause the disintegration of larger particles on freezing.

The chemical action initiated by water is increased, as in most chemical reactions, by heat. It is also greatly assisted by the presence of impurities. In this, vegetation and soil play a very important part. Plants take up nutrients from the soil and ultimately decay. Soil organisms and bacteria work upon this material to produce a cycle of continuously changing chemical compounds. Many of these compounds are soluble in water and are carried downwards by rain through the soil to the rock below. The solution effect on limestone of CO_2 dissolved in water to form dilute carbonic acid and the dramatic caves and swallow holes, mainly the result of this action is well known. In semi-arid climates or in dry seasons, water may move upward through subsoil and soil by capillarity. If evaporation takes place at the surface, the solution will be concentrated and its effect on decomposition enhanced.

In the past emphasis has been placed on the presence of atmospheric carbon dioxide to account for the chemical weathering of limestone (i.e. the solution of calcium carbonate). It is now recognized that CO_2 respired by fine plant roots and largely trapped in the soil, forms an immense reservoir which, because it is in contact with subsurface water for long periods, is of much greater importance in the production of carbonic acid. It is also known that the presence of this acid increases the solubility of water for substances other than limestone — iron compounds for example. Neither has the part which impurities, derived from the soil vegetation complex, play in the subsequent decomposition and solution of many other kinds of rock been appreciated traditionally. Water moves only slowly through subsoil and permeable rock, which thus remains moist for long periods, or is permanently wet. Therefore subsurface chemical decomposition and solution, unlike many other weathering processes, is continuous. *It is now believed by many geomorphologists that in humid areas direct removal by solution is a more important factor in the reduction of the landscape than any other form of denudation.* (Values showing the quantity of solution in some selected rivers are given in Fig. 3.5.)

Both mechanical and chemical weathering can be important in the decay of bare surface rock. Scree slopes, mentioned above as the result of freeze/thaw action, are an important feature of highland Britain. They accumulate most readily in areas not necessarily of the greatest cold, but *where there is sufficient precipitation and the temperature most frequently passes the point which will freeze the water held in the voids within the rock.* In a crystalline rock like granite, water may initially cause surface decomposition of the felspars and so allow moisture to penetrate between and below the quartz crystals. The expansion of water on freezing in this surface layer tends to lever out individual grains, forming a larger depression in which more water can accumulate. The process may continue until a considerable cavity is created which may become sufficiently large for water in it, or in a natural joint plane, to generate enough pressure on freezing to break off the large pieces of rock typical of scree slopes.

Angular pieces of rock are then subjected to further weathering. Sharp edges are attacked from both sides and a process of 'rounding' begins. If accumulation from above ceases, the weathered fragments begin to attract lichen, soil begins to form, and eventually other types of vegetation take over to produce a permanent soil surface. It is important to remember that the accumulation of weathered material on any slope may help to protect subsurface layers from further weathering, until such time as the surface material is removed by the normal processes of denudation. It is also true that most weathered material (other than clay) is permeable and allows the access of water. It follows that the important weathering processes involving chemical action can continue, and in some circumstances rock decomposition may be increased, because a layer of waste material protects it from the drying action of wind and sun.

Nomenclature and mapping

Very rarely do we find areas on the earth's surface that are absolutely level. All landform is composed of slopes – however slight the gradient – sometimes of a very complex nature. Geomorphology is concerned with landform, and so an understanding of the processes which control slopes must form an important element of study. They are rarely completely rectilinear (which means that the slope angle does not change) over their complete length. For convenience different parts of the slope profiles have been given specific names. Unfortunately, as knowledge of slope processes has evolved, these have been altered and added to, and different writers have used different names for the same part of the slope. Fig. 4.1 is an attempted summary of a rather confusing situation.

Fig. 4.1 *Slope nomenclature*

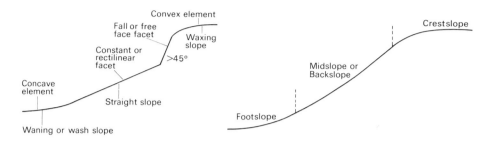

The angle of a slope is always measured in the steepest direction (i.e. perpendicular to the contours). Nearly all slopes consist of a number of distinct parts. Each part is termed a segment. Rectilinear segments are called facets, and those which have a convex or concave profile are called elements. When a facet is at an angle of more than 45° it is known as a free or fall face, the implication being that because of the steep angle any weathered material will be immediately removed. Most slopes consist of a number of segments along the profile (but all those shown in Fig. 4.1 may not necessarily be present in any one slope). The terms footslope, midslope, and crestslope, as their names imply, refer to generalized parts of the profile —in that order — from base level to crest, and each part may contain a number of segments.

The normal representation of altitude by contours is quite inadequate to give more than a very generalized picture of relief. If landform is to be studied scientifically a system of mapping is required to show features too small to be revealed by traditional contouring, and to show in addition the distribution, size, and type of morphological units comprising the area of study. Fig. 4.2a is an example of one method widely accepted. The symbols can be used on their own to make a morphological map or they can be superimposed on an outline contour map to give additional information. It will also be seen that by using symbols it is possible to show what may well be important geomorphological details, such as the difference between a break of slope, in which there is a sharp junction line between two slopes, and a change of slope, where the alteration in angle occurs across a zone on the ground. Fig. 4.2b is a contour map of an area west of Sheffield consisting of two scarps developed in Carboniferous Sandstones, to which morphological mapping symbols have been added. Because arrows indicating steepness can only relate to the line of profile of which they are a part, an indication of the areal extent of land within given ranges of gradient is shown by shading.

Often in the past qualitative description, in which precise measurements were not given, has led the observer to see only what he believed to be there. Previous failure to recognize the extent to which landforms (in Britain for example) are made up of rectilinear slopes is an example of this kind of error. Morphological mapping has helped to provide a useful quantitative cartographic tool in objective geomorphological description.

Fig. 4.2 (a) *Morphological mapping symbols*
(b) *Morphological mapping symbols superimposed on an outline contour map. The areal extent of different gradients is shown by shading*

(a)

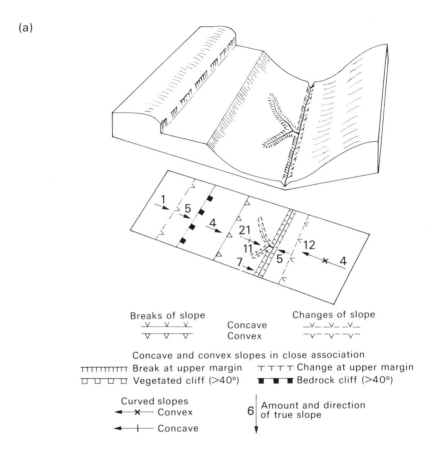

(b)

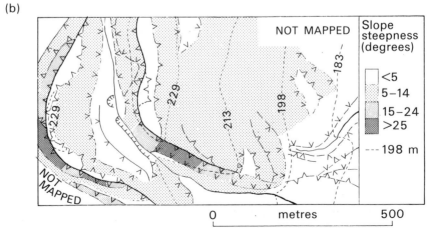

Slope processes

Some slopes, such as actively eroding sea cliffs or the free faces of high valley sides, may consist of bare rock. Or a slope may be formed by a covering of weathered rock resting on bedrock, as we find in some screes. A third type of slope, the most usual in temperate humid areas, consists of bedrock (forming the basal slope), covered by a regolith (i.e. weathered rock, often including a surface layer of soil). It is with slopes of the third type that this section is concerned.

There is as yet no comprehensively accepted body of empirical knowledge to explain slope development. Because the processes which modify slopes under normal conditions work very slowly and the instrumentation designed to measure slope changes is of recent development, there has not yet been time for much observational evidence to accumulate. For these reasons hypotheses concerning slope evolution must be partly speculative, and regarded with great caution. Because of this, much attention is given today to the study of process.

Processes acting upon slopes in humid areas appear to fall into three distinct categories:

1. **Direct removal in solution** due to the passage of water down through the regolith.
2. **Movement on the surface** of the slope as the result of surface wash, raindrop splash, needle ice, and animal movements.
3. Movement of the **surface layer**.

Direct removal depends upon the amount of water available, the temperature, the type of rock, and permeability below the surface, especially of the regolith. The effect of water is dependent upon the amount, intensity, and distribution of annual precipitation, and on the denseness and type of vegetation cover. The speed of most chemical processes increases with temperature. Therefore subsurface chemical change and direct removal in solution might be expected to be at maximum in warmer areas.

Downslope movements as defined in (2) and (3) above are sometimes termed soil creep. They are subdivided here because they may have rather different effects upon the slope profile.

Movement on the surface may be caused directly by the action of falling rain. When a raindrop reaches the ground it bursts on impact. Assuming that the rain is falling vertically (disregarding the wind) the splash of the bursting drop of water will be scattered equally all round the point of impact, carrying fine particles of soil with it. However, an excess of material will tend to be carried downslope because in this direction the force of gravity will be added to the energy translated from the falling drop, whereas upslope the movement will be against gravity. It will be remembered that movement in this way is greatly reduced by the protection afforded by vegetation cover and the organic horizon on the soil surface associated with it.

Animals, including humans, can be a very potent source of downslope movement. Sheep and cattle can be significant on a grassy slope in disturbing surface particles of soil which are pushed downslope by their feet. Large burrowing animals like rabbits may dig out earth, and on a much smaller scale animals like mice, moles, worms, and termites must also make a big contribution to surface movement. In constructing their nests termites may transport considerable quantities of soil to the surface. If the nest is abandoned the void will eventually be filled by subsidence. The surface subsidence hollow will then slowly fill with earth until the ground is once again level. The resulting subsurface irregularity (or involution) may very closely resemble the disturbance caused by the formation of an ice lens under periglacial conditions (p. 83) and these alternative possibilities must always be considered. The small animals are especially important because they usually act over the whole surface of the slope whereas larger burrowing animals are mainly localized.

Frost has a significant effect on the movement of small particles. The white frost which may form on the ground when the ground minimum temperature falls below 0°C is composed of ice crystals. If the temperature falls sufficiently low long thin crystals called needle ice grow perpendicular to the slope upwards from the soil, and may carry particles with them. When they melt the soil particle falls to the ground perpendicular to the horizontal (Fig. 4.3a). In this way every crystal can move a little soil a tiny distance. The total amount of material moved in this way is virtually unquantifiable, but when one considers the huge numbers of ice crystals which form and the number of ground frosts each year, it must be considerable.

If a period of freezing conditions is prolonged the growth of ground ice may take place within the soil. The ground surface freezes first, so the water necessary for growth must come from below. The exact mechanics of this process are not fully understood. Water rises from the unfrozen layers probably partly due to capillarity, and then freezes on to the base of each crystal in such a way as to force it upward. The crystals of pure ice really grow, and in so doing force up the soil above them. Photo 4.1 shows a group of ice crystals dug from the ground in Derbyshire. Crystals of this size — these are about 60 mm long — formed during a prolonged cold spell, can break up the surface material on a slope, and destroy the

Fig. 4.3 *The effect of low temperatures.*
(a) *Needle ice. During a frost small particles are lifted on tiny needles of ice perpendicular to the slope. As the ice melts the particles drop perpendicular to the horizontal, and therefore slightly downslope. Note: these movements are very small. The scale has been exaggerated for clarity*
(b) *Heave. During longer periods of freezing the regolith may freeze in depth — possibly to bedrock. The water content will freeze and expand. Because the ground is frozen overall the only way expansion can take place is upward, perpendicular to the slope. When the ice melts the sideways pressure is relaxed and the regolith tends to subside perpendicular to the horizontal*

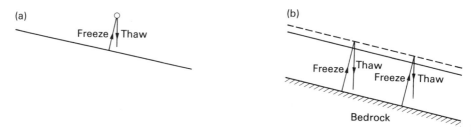

Photo 4.1 *Crystals of ground ice, which grew perpendicular to the slope. (The coin has a diameter of 20 mm.) The dark area at the top is frozen surface soil and vegetation pierced by ice needles. The white layer consists of crystals of pure ice which have grown from below, thrusting the soil layer upwards and in a down-slope direction*
P. James

stabilizing effect of plant roots. Some soil particles may slip downslope when the ice melts, but generally most remain in such a loosened condition that they are readily removed by the next heavy rain.

The amount of surface runoff (or wash) varies with the nature of the soil and subsoil, and with the intensity and/or duration of rainfall. In areas where the ground frequently becomes saturated there is little doubt that surface flow is an important agent in the development of slope profiles. Three different types of flow are recognized:

1. Sheet flow, which takes the form of a continuous sheet of water moving down-slope.
2. Rivulets, which may be so called when flow takes the form of a series of stream-lets, without fixed channels, and which migrate over the surface during a storm, or which are in different places on different occasions.
3. Gullys, which are streamlets which have become canalized into more or less fixed channels which change only slowly, usually as the result of lateral corrasion.

We have already seen how the impact of raindrops can move soil particles. Rain is probably the most effective agent in dislodging them, and surface flow the most important agent of transport. Soil grains disturbed on impact in an area where flow is already taking place can become automatically entrained, and thereafter remain in suspension even when the rate of flow is comparatively slow. Vegetation is very important in protecting the surface during heavy rain, and much less material will normally be removed from a vegetated slope than from a surface of bare earth, such as a ploughed field.

The erosional effectiveness of all kinds of wash varies with the sine of the slope angle. On the steeper slopes velocities will be high and flow will tend to be very turbulent, in which eddies of fast flow (p. 52) will provide considerable erosional power. The lower parts of a slope are generally more affected by wash than those higher up, because water from above, flowing down through the regolith (through-flow) tends to approach the surface at lower levels. It follows that the soil often becomes saturated more quickly after a given period of rainfall on the lower parts of slopes. Consequently, on the same slope runoff may at the same time be active on some areas but not on others.

Certain processes affect not only movement on the surface but also penetrate deep into the regolith. Among these are burrowing animals, especially the larger ones, which in addition to moving surface soil dig into the ground. Eventually the burrows collapse causing a downward movement in depth. Similarly the roots and stems of bushes and trees die and create deep voids through the surface layer which also eventually collapse.

During a prolonged cold spell, water in the upper part of the regolith may freeze and expand. This causes heave. That is, upward movement of the surface in a direction vertical to the line of the slope profile. Because the whole slope is frozen, lateral expansion is not possible and movement can take place only upwards. When the ground thaws again the sideways pressure is relaxed and the return under gravity tends to be perpendicular to the horizontal (Fig. 4.3b) causing a small downslope movement in depth.

Another source of volumetric change is due to wetting and drying. Excess water will cause the regolith to expand and heave may occur with an effect similar to that of freezing. Severe dryness will cause the ground to shrink, and often cracks appear, penetrating the ground deeply. Soil particles from above disturbed by wind or animals may fill the cracks and result in a general downslope movement. Young (1972) reports that in the top 5 cm of sandy loam in northern England he found a shrinkage of between 0·06 and 0·08 per cent for every 1 per cent change in the moisture content of the soil (i.e. moisture as a percentage of the dry weight of soil). In this location there was an annual moisture change of 35 per cent giving a change in linear dimensions of between 2 and 3 per cent.

The measurement of slope processes, by which is meant the measurement of the weathering and movement of material on slopes, is one of the more difficult aspects of geomorphological study. The difficulties are of two kinds. Firstly, there is the problem that the span of time required to detect any measurable amount of soil creep is long, and experiments have to be thought of in terms of years rather than weeks. And secondly, there is the practical consideration of the physical difficulties in finding suitable sites where instruments can be placed in positions reasonably secure from interference by people or animals. If these problems can be overcome there is much interesting real research to be done, because to date so little has been published. Figs. 4.4 and 4.5 indicate some rates of soil creep and surface wash that have been measured in areas of different climate.

Fig. 4.4 *Observed rates of soil creep*

Source	Method	Climate	Movement of surface or upper 5 cm, mm/yr	Volumetric movement, cm³/cm/yr
Young, 1960, 1962	Surface and buried pegs	Temperate	1–2	0·5
Everett, 1963	Buried plates	Temperate	1	—
Kirkby, 1964, 1967	Surface and buried pegs	Temperate	1–2	2·1
Slaymaker, 1967	Buried pegs	Temperate	—	2·8
Owens, 1969	Buried tubes	Temperate	11	3·2
Williams, 1969	{ Buried pegs and tilt-bars	Warm temperate / Savanna	— / —	1·9–3·2 / 4·4–7·3
Schumm, 1964	Surface pegs	Semi-arid	6–12	—
Leopold et al., 1966	Surface pegs	Semi-arid	5	4·9

Source: Young (1972), p. 56

Fig. 4.5 *Observed rates of surface wash*

Source	Method	Climate	Ground lowering mm/yr	Volumetric movement, cm³/cm/yr
Young, 1960	Traps	Temperate	—	0·08
Starkel, 1962	Traps	Temperate	<0·005	—
Gerlach, 1963	Traps	Temperate	<0·005	—
Gerlach, 1967	Traps	Temperate	0·03	—
Kirkby, 1967	Traps	Temperate	—	0·09
Smith and Stamey, 1965	Experimental plots	Temperate	0·01–0·06	—
Soons and Rayner, 1968	Traps	Temperate montane	0·01	—
Williams, 1969	Traps	Warm temperate	0·05–0·10	—
Gabert, 1964	Traps	Mediterranean	0·09	—
Schumm, 1964	Stakes	Semi-arid	2·00	—
Leopold et al., 1966	Stakes	Semi-arid	6·40–8·20	—
Williams, 1968, 1969	Traps	Savanna	0·039	—
Rougerie, 1956	Stakes	Rainforest	5·0–15·0	—

Source: Young (1972), p. 70

Slope forms

Some slope forms are simple, like those illustrated in Fig. 4.1. An uncomplicated profile generally occurs when the underlying rock is homogeneous, although the reverse is by no means true. Complex slope profiles may reflect differences in erodibility related to changes in rock type, or they may be due to past changes in

Fig. 4.6 *More complicated slope profiles, with unimpeded removal.*
(a) *Resistant bands of rock, each forming a free-face facet and acting as base level for the element above it, complicate the profile*
(b) *Slope profile complicated by erosional history. Rejuvenation has left old terraces forming part of the upper slope. In this case there is unimpeded removal only from the bottom of the slope*

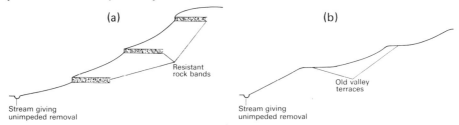

climate or base level (Fig. 4.6). The bottom end of a simple slope constitutes its base level. (If the profile is complicated, for example by bands of harder material, each band of resistant rock acts as the base level for the slope above it.) If base level is formed by a stream which transports any material moved down the slope there is said to be unimpeded removal. Unimpeded removal can also be provided by the sea when material weathered from cliffs is removed by waves.

There are a few nearly vertical slopes consisting of hard unjointed rock (mostly sea cliffs). These maintain a very steep gradient because the granules of which they are made are cohesive. That is, the granules have become fastened to each other by some natural form of cement, or have been under such great pressure that a chemical bond has formed between them. But large outcrops of hard unjointed rock are rare. All other kinds of slope depend for their stability (i.e. their ability to stand at a given angle) on three factors:

1. **Friction** between particles, large or small, ranging in size from the large blocks of a well-jointed rock to finely comminuted fragments. The amount of friction depends upon pressure and the roughness of the surfaces touching each other.
2. The way in which the particles are **interlocked**. Generally the larger the particles the more effective this will be. For this reason a pile of coarse gravel will stand at a much steeper angle than a heap of *dry* sand. The most effective interlocking occurs in some massively jointed rocks which may provide a vertical face.
3. The **matting effect** of the roots of a close vegetational surface cover.

A scree provides a common example of a slope consisting of large, rather rough lumps of rock. Because of this screes generally stand at a steep angle. However, most slopes in temperate areas consist of a regolith, that is, a layer of decomposed rock, overlying bedrock, with a surface of soil and vegetation. (Clay is a rather special case and is considered below.) When the soil and subsoil are moist (the most usual condition) the water adheres to the surfaces of individual grains, but does not fill the voids between one grain and another. In simple terms, when these conditions exist, chemical bonds are made consisting of small residual electric charges. In the case of quartz (SiO_2) the oxygen possesses a residual negative charge. In water (H_2O), the hydrogen has a residual positive charge. If two sand grains touch in the presence of water an attraction will exist between the hydrogen in the molecules of water and the oxygen in the molecules of silicon dioxide, which helps the particles to cohere, or stick together. (This is why sand castles have to be made in damp sand.) Therefore some moisture helps to maintain slope stability.

If the ground becomes saturated the situation changes. Imagine weighing a stone tied to a piece of string attached to a spring balance. Lower the stone into a bucket of water and the balance will record a loss of weight by the stone equivalent to the amount of water displaced. When the ground is saturated each particle becomes immersed in water and loses the equivalent of the weight of water it displaces. It becomes less stable, because the pressure between one particle and another, and therefore the friction between them, is reduced. This is known as pore water pressure. In places where gradients are steep and a very thick regolith is composed of a high proportion of fine rock particles – conditions which are found, for example, in the Himalayan foothills, in parts of the Andes, and on a small scale in some places in Britain, the addition of a large quantity of water may result in sudden rapid earth flows or earth avalanches (Photo 4.2).

A slope is said to be stable when the surface is not subject to sudden downslope movements. Signs of instability are bare patches of soil or rock on an otherwise vegetated slope, indicating where a landslip or flow has taken place. Terracettes are also an indication that soil creep is rapid and the surface potentially unstable.

Vegetation aids slope stability in two ways. We have seen above how it partly protects the soil directly from erosion. The matting effect of roots is also important in preventing the surface from slipping, especially under saturated conditions. But even with a thick vegetation cover the processes of denudation, although reduced,

Photo 4.2 *Mudflow, Sulby Glen, Isle of Man. Terracettes indicate the unstable nature of the main slope. In the centre is the lobe of the mudflow, which must be of more recent origin than the footpath which passes beneath it*

still continue. It sometimes happens that, with unimpeded removal, part of a slope will continue to steepen beyond the natural angle of rest for the material of which it is composed, held in position by the strength of the roots of the vegetation it supports. If the ground now becomes saturated due to heavy rain two things occur. Pore water pressure reduces the friction between particles, making stability even more critical, and at the same time the weight of the surface layers is increased greatly by the weight of the water which has been absorbed by them. The result may be a slip or rapid flow downslope until the angle of the over-steepened section has been sufficiently reduced to achieve a new stable gradient.

Clay is different from other sedimentary rocks because its particle size is so small (< 0.002 mm). Each particle holds a film of water around it so firmly attached by surface tension that it is unavailable to plants. The total surface area available to hold water is very large. As the clay takes up water it increases in volume and this may well be an additional factor in causing landslips. When a clay has absorbed all the water it can hold, although porous it becomes impervious. Clays become cohesive under pressure, so cohesion will increase with depth from the surface. Some clays now exposed which have been under pressure from previously overlying strata may retain a large measure of cohesiveness even at the surface. These are sometimes called consolidated clays.

We have seen that, apart from sudden, catastrophic movements like landslips, it is believed the major processes affecting slope form are physical and chemical reactions, solution, soil creep, and wash. Of these, solution is coming to be regarded as the most important agent of denudation. The water containing dissolved impurities also reacts in a complex way with the regolith and underlying bedrock (some of which may be removed in solution), leaving behind altered products which may not be less in bulk than the original material. Until the altered products are themselves removed, surface lowering will not take place. Nevertheless, mass is being removed, and the amount of the measured solution load of streams shows the importance of the process. Some quantitative evidence has been accumulated of the individual processes operating on natural slopes, but much more research is needed in different areas and under varying conditions. The door is open wide here for many people to contribute. Provided the experiment is carefully controlled and accurate measurements meticulously taken, the possibility of research is not limited to the field of higher education.

The interaction of all the processes operating upon a slope appears very complex, and an accepted body of theory is only just beginning to emerge. Perhaps one of the best ways to approach the slope problem is to consider it as a working system. The application of systems theory to geomorphology was first proposed by Chorley in 1962, and a systems approach to the whole of physical geography was developed later by Chorley and Kennedy in 1971. They describe a system as 'a structured set of objects and/or attributes'. Systems are of three kinds: isolated, closed, and open. Isolated systems are those enclosed within boundaries preventing the passage of both energy and mass. It is difficult to imagine this occurring naturally, but it is a useful concept because it enables study of the effect of external variables on a particular system. Closed systems are those in which energy alone may be transmitted across its boundaries. The global mechanisms of weather may be regarded

Fig. 4.7 *A valley slope as a physical open system. Open arrows show movement of energy and material across the boundaries of the system. Solid arrows show movements within the system. The load carried by the wind may be fine dust particles, but often more importantly in humid areas water vapour from the ground and vegetation*

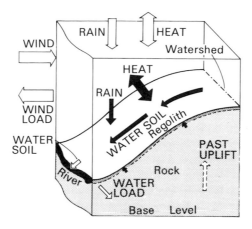

as a huge closed system operated by inputs of energy from the sun and exports of energy radiated back into space. In an open system there is an exchange of energy and mass across the boundary. An individual slope, or part of one, may be regarded as an open system, in which there are inputs of energy in the form of solar heat, rain, wind, and past uplift, and mass from above. Exports of energy out of the system take the form of stream flow and radiation. Mass is exported as the stream's load and as airborne material (including water vapour) (Fig. 4.7).

The slope itself may be considered as bedrock with a mantle of regolith continually moving downwards as a result of past and present energy inputs. The stream acts as base level for the slope and provides the means of transporting water and soil brought down the slope across the boundary out of the system. The important thing is not to regard a slope as an inert mass, but as a dynamic moving natural mechanism with constant complex interactions between the force of gravity, rain, heat, wind, and the vegetation and rock of which it is made.

It is emphasized again that slope processes and their effects are not yet fully understood. Bearing this in mind Fig. 4.8 is a hypothetical slope model devised by Dalrymple, Blong, and Conacher to represent diagrammatically the classification of slope form they found in a temperate humid part of New Zealand. The nine units they describe, and the processes associated with them, are not comprehensive. Other units may be necessary in different conditions. Nor would all the units necessarily be found on every slope. For example, the free (or fall) face is generally much less common than the other units. It is also to be anticipated that in a complex profile a specific unit may be repeated one or more times. It will be observed that the terms used to define slope units involve processes as well as slope form. This is inevitable, because the relationship between form and process is not properly understood. We cannot even be sure if process causes form or if the reverse is true, or even whether both act independently. However, it remains a useful quantitative tool for comparison.

The model (Fig. 4.8) may conveniently be divided into three sections for examination of form and process. Note that the processes mentioned are those which are thought dominant in each unit, and that the units (other than the fall face) and the suggested processes merge slowly into each other.

1. **The convex crestslope** (units 1-3). Here the main processes seem to be surface wash and creep, and removal by subsurface soil water. The gradient is so slight in unit 1 that nearly all movement is through the downward action of soil water. As the angle begins to steepen on the seepage slope there is a larger component of downslope movement, but this is still mainly confined to subsurface water. On unit 3 the dominant process is creep; with growing steepness the speed of movement increases, the slope becomes potentially unstable, and terracettes form.

2. **The mid or backslope** (units 4 and 5). Units 4 and 5 are characterized by steepness and therefore rapid movement. As its name implies, removal is direct and immediate from the fall face. On the transportational midslope the slower agents of denudation are at work, but so also are the more rapid ones of flow and slip, producing a very unstable unit in the profile. On unit 5 would normally be found bare patches of soil or rock, and frequently the slope would be rectilinear in form.

3. **The footslope** (units 6 and 7). This is generally an area of stability in which solution and the slow downslope processes are dominant. In a temperate humid area there would be no bare soil or rock, nor would there be obvious signs of movement other than (possibly) soil creep. The steeper upper part of the footslope appears rectilinear, becoming concave lower down. The concave part, the bottom end of unit 6 and unit 7, is particularly interesting, because it is shown on the diagram as an area of deposition. Footslopes of this kind are very common and it is not unusual to find them so described. Yet a little thought will show that if this process were to continue the concave element would steadily grow and eventually consume the backslope. If this is not to occur, direct removal by subsurface water, plus transport downslope to the stream and possibly some movement downvalley must equal the amount of deposition from above. This is what appears to happen, but how is not clear and requires much more research. It has an important bearing on the problem of equilibrium discussed below (p. 74).

Solution is important not only for the quantities involved but because removal is direct and takes place over the whole slope. If we disregard all processes other than solution, provided removal was equal over the whole profile, a rectilinear slope would remain at a constant angle and retreat parallel to itself. We know this does not necessarily happen over a natural slope because other processes are at work which do not equally affect the whole slope: for example, the lower area gets saturated first and surface wash is therefore more effective here than higher up. Nevertheless, in temperate humid areas it may well be that direct removal in solution may be a major mechanism in the preservation of the many rectilinear slope facets existing naturally.

Fig. 4.8 *Diagrammatic representation of a nine unit land surface model*

1. *Interfluve* $0° – 1°$
2. *Seepage slope* $2° – 4°$
3. *Convex creep slope*
4. *Fall face* $> 45°$
5. *Transportational midslope* $26° – 35°$

6. *Colluvial footslope*
7. *Alluvial toeslope* $0° – 4°$
8. *Channel wall*
9. *Channel bed*

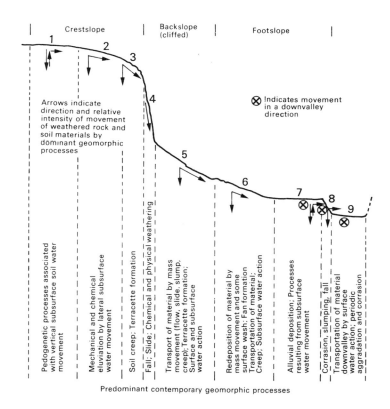

Slope analysis

Slope analysis must be the result of measurement, never merely of inspection. The careful survey of slope angles resulting in the construction of profiles similar to those in Fig. 4.8 is the beginning of analysis, because from these quantitative measurements comparisons can be made between slopes. A few techniques are suggested here, but the method of representing and analysing data will vary very much with the problem involved and many additional methods could be used.

Fig. 4.9 shows two ways of the simple representation of three slope segments – convex, rectilinear, and concave – drawn as the result of the survey of a series of profiles. Fig. 4.9a describes the changing proportions down one small valley side from source to base level. The diagram is accurate only at the places where each of

Fig. 4.9 *Methods of representing slope composition. For explanation see text*

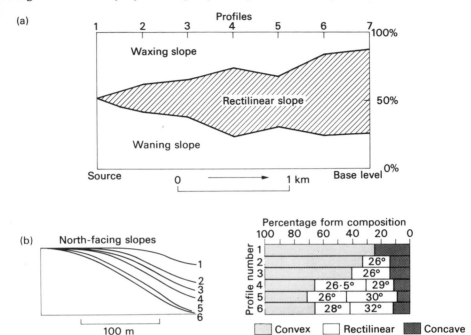

the seven equally spaced profiles was surveyed. Because no information was available between profiles, known points of change were connected with straight lines in order to give the effect of continuity. Fig. 4.9b shows a slightly different way of conveying the same sort of information. Additional detail has been given to show the angles of the rectilinear proportion of the profiles. The actual lengths of the slopes are shown to scale, but not the spacing between them, nor the apparent exaggeration of the vertical scale.

The relationship between slope steepness and geology was demonstrated by Doornkamp and King (1971) by grouping the measured angles into classes and representing the proportions in each class by a histogram (Fig. 4.10). This is a quantitative descriptive method of comparison. If it is desired to determine whether the observed differences are statistically significant simple inferential tests are required. For example, the significance of the difference between mean slope angle developed on two kinds of bedrock could be determined by using the Mann Whitney U-test (*Science in Geography 4: Data Use and Interpretation*, p. 38). Or a significant difference in the frequency of slopes of a given class occurring on a number of geological outcrops (or of aspects) could be estimated from the chi-squared test (*S.I.G. 4*, p. 49). Note that percentage values cannot be used with the chi-squared test.

Correlation is a useful form of analysis, because a correlation coefficient is both a descriptive numerical index of the degree of association between two sets of paired variables and also a measure of the level of statistical significance of the degree of association. A distribution-free method of correlation is desirable for this

Fig. 4.10 *Histograms showing slope steepness on three different bedrocks in the Weymouth lowland*

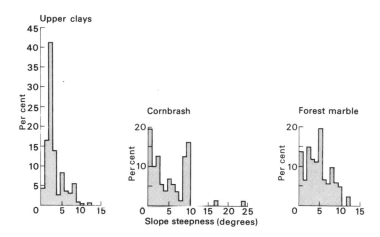

purpose, because often the number of sample values obtainable is small and the distribution cannot be assumed to be normal. Kendall's (tau) or Spearman's rank correlation coefficients are the most suitable for this purpose. (*S.I.G.4*, pp. 4, 77, 85.) Most pre-programmed calculators use Pearson's product moment formula. (A note on this will be found in *S.I.G.4*, Appendix 8a.)

Correlations might be used, for example, to establish the degree of association between a series of mean angles of waning slopes and those of the long profile of a stream at a series of randomly chosen cross-sections down one side of a valley. (The length of the long profile used at each point is arbitrary. It might be 50 m above and below the section/stream junction.) A high coefficient of correlation might be an indication that the lower slope at least, and the stream, were close to equilibrium.

There are many slope variables which are worth correlating. Some are shown in Fig. 4.11. Examples of others could include mean slope angle and slope length, percentage of rectilinear slope on either side of valley cross-profiles (to detect asymmetrical development), slope angle and altitude, distance upslope and soil moisture content, and many more. The result of a number of correlations referring to the same area is usually best displayed in the form of a matrix (Fig. 4.11). All the coefficients obtained may be entered as descriptive indexes, or, depending on the way the information is to be used, only those which are significant. If information is in the form of frequencies and either variable is measured on the nominal scale, the chi-squared contingency coefficient may be used as a kind of correlation coefficient – for example, to test whether there exists an association between aspect and slope angle (*S.I.G.4*, p. 88).

Care should be taken in choosing variables for correlation so that autocorrelation is avoided, that is, a situation in which one variable may be directly dependent upon another. For example, the correlation of slope length and altitude might display autocorrelation.

Fig. 4.11 *Correlation coefficients for selected attributes of a small river valley displayed in the form of a matrix. Variables are as follows:*

1. *Slope length*
2. *Angle of stream long profile*
3. *Percentage of soil moisture*
4. *Altitude*
5. *Soil depth*
6. *Mean angle of waning slope*
7. *Percentage of waning slope*
8. *Percentage of rectilinear slope*
9. *Mean soil particle size*
10. *Soil pH value*

Variables 5, 9, and 10 may consist of an average value for the whole profile, or more meaningfully relate to particular slope segments.

   A full correlation matrix is always symmetrical and for this reason normally only half is given. It is sometimes easier to identify whether the correlation is negative or positive if positive coefficients are placed on one side of the diagonal and negative coefficients on the other

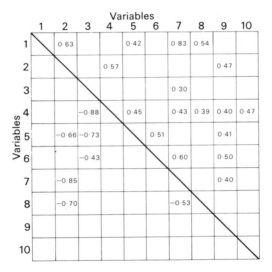

Dynamic equilibrium

Most modern writers of geomorphology have at some time considered the problem of dynamic equilibrium, a concept first foreshadowed by Gilbert as long ago as 1877. The term equilibrium has been previously used (pp. 12, 55, 70) to denote a state of balance, in which erosion and transport on slopes is matched by the capacity of the stream to remove waste material, and in which the lowering of the valley slopes matches the rate at which the streams are downcutting. Dynamic equilibrium may in theory be applied to the landform of a drainage basin, or to a single hillslope, or a short reach of one stream. Let us consider it here in relation to the simple slope profile shown in Fig. 4.12, a type often found in a temperate humid climate.

   We know that the processes of solution, wash, and soil creep are continually removing material from the slope, some directly by solution, but some also downhill on the surface. If this is happening, and we know it does happen because it has been

Fig. 4.12 *A simple hypothetical slope profile commonly found in temperate humid areas. A is the base of the rectilinear facet*

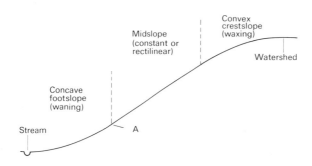

measured, either the soil is accumulating at the bottom of the slope, or it is transported into the stream and removed. It is not always easy to know which of these alternatives to accept because the process is so slow. But on the evidence that we have, it seems reasonable to suppose there is no accumulation on many natural footslopes, and therefore waste material is reaching the stream. We know that material in solution is being removed directly from the whole length of the slope. The altitude of the profile is therefore slowly being lowered. In order to maintain the waning slope the stream channel must be reduced at exactly the same rate. If the stream were downcutting more quickly the waning slope would disappear; if less quickly deposition would occur. When all these factors are in fine balance there is said to be a state of dynamic equilibrium. Of course minor fluctuations take place from time to time, for example, minor changes in base level, or small increases in runoff, both leading to an increase in energy (and therefore erosion capacity) of the stream. These small fluctuations are generally described as 'self regulating' through the operation of negative feedback (p. 55).

We are still left with real problems in what may, for the want of a better term, be called slope dynamics. It is probable that relatively little surface material moves downward in the crestslope section, partly because the gradient of most of it is not very steep, and partly because there is more surface wash on the lower part of the profile. The steepest part is usually the midslope, often rectilinear in nature, and made so in Fig. 4.12. It seems logical that the rate of transport would be at a maximum on the steeper gradients, in this case down the midslope, but strangely enough there is some evidence that this is not so. Young in 1960 discovered little difference in the rate of creep on 26° and 7° slopes. But much more information is necessary before it is possible to be sure.

The existence of many waning slopes with very slight gradients is difficult to explain, because we are not yet sure exactly what processes are at work on them. It is unlikely that removal (other than by solution) can be as effective at the very gentle angle of the waning slope, where the influence of gravity is greatly reduced, as on the steeper midslope. There is evidence that particles become more finely comminuted in their journey downslope, and thus more easily removed by wash. We have seen that the soil of the footslope is likely to become saturated more often than on the upper slope, and movement by surface wash will therefore be more frequent and effective in this area. On the other hand Young found creep nearly ten times more effective than wash where there was close vegetation. The increased frequency with which wash occurs on the footslope may be an important factor in surface transport. Fig. 4.8 indicates that there is also a downvalley component of movement near the base of the profile due to the downstream inclination of the valley floor, and wash may be an important causal factor for movement in this direction. The fact remains that waning slopes exist in nature. We have to regard them either as areas of deposition, and therefore of recent origin, or as areas in balance and able to transport all the material received from above — either into the stream or downvalley, or both. Nor should the possibility be overlooked that they could be relict features, formed perhaps in a periglacial climate, or when river discharge was greater. They may not now be in equilibrium but may be slowly being modified under present conditions. Much more research is required before we shall know the answers to many of our questions concerning the way slope mechanisms work.

Chorley and Kennedy (1971) have suggested that a large number of significant correlations between various aspects of the physical landscape are an indication of dynamic equilibrium. If a stream provides unimpeded removal at the base of a rectilinear slope (point A on Fig. 4.12), it is possible to envisage a state of approximate equilibrium, with minor fluctuation of stream discharge corrected by negative feedback, provided that the rate of erosion of the stream bed matches the reduction of the slope. But if the stream is at the base of a waning slope, then unless there is removal downslope (and possibly downvalley) of all the material received from above, equilibrium cannot exist.

Consolidation

1. Why should a large number of variables, significantly correlated, indicate a state of dynamic equilibrium?
2. Describe the processes thought to be most important operating on slopes in temperate humid areas. Include a detailed account of the mechanism of soil creep. Where possible give a local example to illustrate each point.
3. Measure the profile of a real slope. To what extent do the processes described in (2) appear to be evident? Are there any anomalies? If so can you account for them?
4. What problems are involved in explaining the presence in temperate humid conditions of the many waning slopes of gentle gradient that we see around us?
5. Using the symbols given in Fig. 4.2 attempt a simple morphological map of a small local area to show landform detail which is not revealed by contours on an O.S. map.
6. Try to measure both surface wash and soil movement on a local slope by devising your own instruments. Some suggestions are made in this chapter, but there is great scope for the exercise of ingenuity.
7. What makes a slope unstable? How do you recognize this? Can you give a local example?
8. From the correlation coefficients displayed in Fig. 4.11 what tentative conclusions can you suggest concerning the area to which they refer?

Chapter 5

Periglacial and glacial landforms

Periglacial conditions

Landforms sculptured by ice or the result of glacial and fluvio-glacial deposition have long been studied. Of much more recent origin is interest in the effect of periglacial processes. Periglacial geomorphology is a part of geography which previously has received far less attention than its importance warrants. Periglacial conditions, past or present, still actively affect more than one-fifth of the land surface of the world. Ice sheets and glaciers during the Pleistocene have greatly modified the areas affected (Fig 5.1 shows the extent of the ice sheet in Europe), but periglacial conditions, marginal to the ice sheets, and generally resulting in less dramatic effects than glaciation, are equally important if we are to interpret landforms existing in many temperate areas today. Some landforms in Britain, for example, can only be explained in terms of the kind of processes prevalent in the climatic conditions of the Pleistocene, whatever their subsequent and comparatively short erosional history has been.

A knowledge of the physical processes active in cold climates is also becoming more important as many mineral and other resources are being currently discovered in the far north. The exploitation of the Alaskan oilfield and the mineral resources

Fig. 5.1 *Limits of the Pleistocene ice sheets in Europe at maximum*

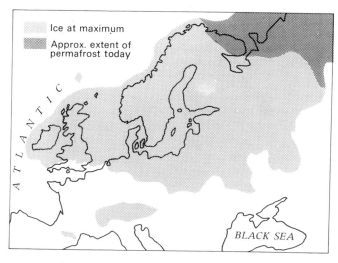

Ice at maximum

Approx. extent of permafrost today

ATLANTIC

BLACK SEA

of the Canadian Shield are examples. Building construction presents great problems in areas of permafrost, a matter of some importance when we realize that continuous or discontinuous permafrost underlies 47 per cent of the U.S.S.R., probably 50 per cent of Canada and Alaska, and obviously most of Greenland and Antarctica. Towards an understanding of these problems the geomorphologist has a real contribution to make.

The word periglacial literally means 'around' or 'near' ice. It refers to areas at high altitudes or close to an ice sheet which experience very cold conditions and where freeze/thaw processes are dominant. This is the only possible definition, because there are considerable differences between the present climates of areas which may be defined as periglacial. With reference to the Pleistocene the term 'periglacial climate' means almost as little as a 'monsoon climate' today, and has tended only to conceal the variations between one area and another. The climate in southern Britain, for example, during the Pleistocene glaciation was of the maritime type associated with less cold and more precipitation than the colder, much drier conditions of central Russia. The cold dry winds of the continental interior explain the distribution of huge amounts of loess, virtually unknown in Britain, in northern Europe, central Asia, and north China.

To compare present periglacial climates of the Arctic with those of previous glacial periods is not possible, because periglacial areas in the Pleistocene occurred at lower latitudes, where the effect of insolation was greater. Surface melt would have been quicker and probably have lasted for a longer period of the year. The essential characteristic of periglacial conditions, as they affect landform, is the dominance of freeze/thaw action and the low temperatures which cause the ground to become frozen in depth. Also important for the mechanical weathering of rocks and for mass movement under gravity is the number of times the temperature crosses freezing-point annually, and the extent of the temperature range. Both are likely to have been greater in lower latitudes, and therefore more effective in the Pleistocene than they are in present periglacial areas. Periglacial conditions more closely resembling those of the Pleistocene are probably to be found today marginal to the high snow of the Himalayas or the Alps.

To be effective freeze/thaw action requires water, though not necessarily in the form of rain. It can be the result of snow melt, which is probably important in breaking up larger pieces of rock, or ground or soil water. Dew or melted frost can provide sufficient moisture for the disintegration of small surface grains, or help to enlarge hairline cracks. Little research has so far been done on chemical weathering in cold conditions, but there exists some evidence that it still remains important, especially associated with freeze/thaw action. For example, it is known that carbon dioxide is more soluble in water at low temperatures, and therefore higher than average concentrations of carbonic acid are likely to be found in snow melt. It is also possible that contractions of the rock due to very low temperatures assist disintegration.

One important result of a periglacial climate may be the formation of permafrost, or permanently frozen ground. (Though this does not necessarily follow.) Permafrost is found in places where there is a negative heat balance at the earth's surface, that is, where more heat is lost from the ground by conduction into the atmosphere and by radiation than is received from the sun or the interior. To some extent, surface ice, and thick snow act as insulators and will help to prevent the ground's cooling. The greatest depths of permafrost are therefore found in central continental areas like Siberia where extreme cold and low precipitation have existed for long periods of time. Here the thin snow cover reflects a high proportion of incoming short-wave solar energy, whilst the clear skies permit the long-wavelength radiation of heat directly into space, with the result that the ground is frozen in places over 600 m deep. The depth attained in Britain is not known, but even during glacial periods, which were of shorter duration in Britain, the climate was less cold and more humid than in central Asia, and permafrost could never have approached present Siberian depths. In the extreme north, however, permanently frozen rock has been found in a colliery in Spitzbergen about 320 m below the surface. Whether permafrost in these areas is a result of past conditions, or of present climate, or possibly a combination of the two, is a matter of speculation. The last possibility is probably true of most places. (There is evidence at Port Nelson, Manitoba, that permafrost has formed under present climatic conditions and will continue to be maintained to a depth of about 10 m.) The thickness of permafrost in Russia suggests that it must in part be a relict feature.

Permafrost is commonly recognized as of three kinds:
1. **Continuous**. This is found in areas cold enough for permafrost to form today.
2. **Discontinuous**. Where there are small scattered *unfrozen* areas.
3. **Sporadic**. Small scattered *frozen* areas. Probably the result of a previous colder climate and therefore relict features.

Transport

Streams
Streams differ from those of temperate areas through being seasonal. During the cold part of the year the water is frozen. When the thaw comes in the short summer, melting is rapid. Valleys carry a great volume of water for a brief period and large quantities of debris can be transported. Braided stream beds are thus very common.

Solifluction
Solifluction is the movement downhill of a surface layer saturated with water, and probably the most important process of transport on slopes in periglacial conditions. It occurs in permafrost areas when the ground surface melts. This may be seasonal and last through the short summer, in which case the unfrozen layer may be a metre or more in depth. At the beginning and end of summer, freeze/thaw often occurs diurnally. In the spring, daily surface thawing may initially be limited to a depth of a centimetre or so, freezing again at night. As autumn sets in the reverse is the case, with night frost freezing a thin crust at the surface which later melts during the day.

The depth of ground subject to melting is called the active layer (in which vegetation may root). Frozen ground may contain up to 80 per cent of ice, and more water may also be provided by snow melt or rain. If the latter, it has the

Photo 5.1 *A striking visual example of solifluction at Prawle in south Devon. The present cliff is of 'head' (a mixture of mud and angular rock fragments) that has flowed down from above between the rocky outcrops of a previous cliff line*

added function of accelerating the thaw. Because water cannot penetrate frozen ground, the active layer becomes quickly saturated. Friction between the soil particles is reduced as pore water pressure in the soil rises (p. 67). At the same time the weight of the active layer is increased by the amount of water absorbed. The resulting mud, often mixed with shattered fragments of rock, may move quite rapidly downhill under the action of gravity, possibly up to several metres in a few days (Photo 5.1). Speed of movement depends on the steepness of the slope, the proportion of water (the more liquid the mud the faster it can flow), and the type and thickness of the vegetation cover. (Note that solifluction is not confined to areas of permafrost, but seems most effective in this association. It can occur under extreme conditions in all types of climate, other than very arid or permanently frozen areas.)

The terms developed to describe deposits derived from periglacial processes are numerous and tend to be confusing. Congelifluction has been proposed as a more precisely defined term than solifluction. In Britain solifluction deposits have long been referred to as head. The angular fragments of frost-shattered rock found in screes or possibly in head are sometimes called congelifractate. Disturbance and movement by freeze/thaw action associated with water saturation has been termed congeliturbate. Cryoturbation may be applied to any kind of disturbance — such as heave — due to freezing.

Nivation

Nivation is a method of erosion and downhill transport which may occur wherever there is a snow patch. It is, however, most effective in periglacial areas. It is a term used to describe erosion associated with seasonal (occasionally with semi-permanent) patches of snow.

Normally a snow sheet will protect the surface from erosion but, where a *patch* of snow remains, erosion will begin during periods of freeze/thaw around its lower edges. During the daytime some of the snow will melt and the meltwater either soak into the surface below the snowpatch or run downslope. If the surface is already saturated, as will frequently be the case, all the meltwater will run downhill. During a night-time frost, the saturated ground will refreeze. In this way frost will be continually reducing the size of the surface material, while the meltwater provides a method of downslope transport for finely comminuted particles, and ensures the presence of water in the soil for the next frost to be effective. Removal of material in this way may enlarge the hollow enough for sufficient snow to accumulate to last through the summer, so providing a continuous supply of water and greatly increasing transport and therefore effective erosion (Photo 5.2). If permafrost is present, erosion may be increased because the surface refreezes more readily, and solifluction may assist in the removal of debris since meltwater is unable to soak into the ground and all is available for transport. As the snow melts, the patch

Photo 5.2 *Large nivation hollow, Glen Feshie, Scotland. The photograph was taken at the end of August when the area of the snow patch was at its minimum. The unvegetated ground indicates the extent of the snow at the beginning of summer. The steep wall at the back of the hollow is typical*

decreases in size, continuously exposing new ground surface for erosion. In this way the process of nivation associated with a single snow patch may spread over quite a large area.

Wind

Wind may be an important transporting agent in periglacial areas, which characteristically have a dry climate. They frequently exist close to ice sheets, or occupy areas recently glaciated. Here large quantities of deposits in glacial outwash plains often provide reservoirs of dry sand and dust, lacking cohesion and susceptible to movement by wind. Under the dominant westerly winds huge quantities of fine dust, called loess, were deposited along the Pleistocene ice margins from northern Europe to north China (where it has been found up to 300 m in depth). Much sand was also moved in this way. In central Poland nearly all sands are aeolian deposits of Pleistocene age. (It is possible to differentiate wind blown sand because transport in the air by saltation (page 105) gives the grains a matt surface.)

Common periglacial phenomena

Frost heaving

We have already seen that some frost heave may take place during cold spells in temperate climates, and that needle ice also plays a minor role. Under ideal conditions it might be expected that heave would play an important part in surface movement. However, it has been shown that the freezing of interstitial soil water could only cause a 10 per cent expansion at maximum, whereas in very cold areas heaving requiring a 100 per cent expansion has been observed.

The mechanism appears to be as follows. The size of particles of which the

Fig. 5.2 *Soils that are susceptible to frost heaving. Notice that as the proportion of the soil with a large grain size (i.e. sand) increases, ice lenses cannot form and frost heave will not occur*

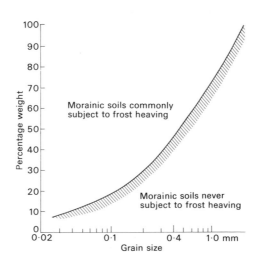

ground is composed is very important. Heaving occurs most readily in silts, occasionally in clays, but rarely in coarse sands. Fig. 5.2 shows the composition of soils by grain size susceptible to frost heave. The layer of ground near the surface is first reduced to a temperature below 0° C and the water held within it is frozen. The ice in the base of the frozen layer attracts water from below, which freezes onto it molecule by molecule, rather like the growth of ice crystals in clouds. Thus a thickening layer of clear ice is formed fed by water from below. (It must be from below because the frozen surface prevents downward movement from above.) Capillarity may play some part in the upward movement of water and the water may be drawn upwards by tension, i.e. in molecular chains. Fine-grained materials are necessary to maintain the very thin columns of water required, which may explain why grain size is critical for heaving to occur. But the process is not yet fully understood. Particularly obscure is the mechanism whereby molecules of water force themselves between the underside of the ice layer and the soil beneath it, exerting sufficient pressure in some cases to lift many metres of overlying material as the layer thickens.

The layer of clear ice (called segregated ice) will continue to grow while freezing continues, and while the flow of groundwater lasts, or until the ground below it also freezes and the supply is cut off. The ice layer generally forms in a lens shape parallel to the surface, and it is the thickening of the lens which causes surface heave. Such lenses have been found in Alaska 4 m thick. Near the Yenesi river in the U.S.S.R. in a shaft 30 m deep a series of lenses of segregated ice was found with a total thickness of 7·6 m. If the ice melts, slumping occurs and hummocky irregularities appear at the surface.

Stones within the active layer may be lifted by ice action. They are a poorer insulating medium than soil and the area immediately beneath each stone tends to be colder than its surroundings. When low temperatures occur little ice lenses tend first to form below stones, forcing them upwards. When there is a period of thaw and the lenses melt, fine material may be washed into the spaces below the stones and prevent their falling back. If this happens sufficiently often the stones may appear on the surface. On flat ground, stone circles may form. First a circular mound is created by the formation of an ice lens, and any stones pushed to the surface by the method described above tend to roll down to the edge to make a circle (Fig. 5.3). During the summer thaw the mound tends to flatten, to be reestablished the following winter: a process which accelerates the movement of

Fig. 5.3 *The formation of stone circles:* (a) *stony silt deposit*, (b) *a mound is formed as a result of frost heave*, (c) *stones are pushed to the surface and move downward to form a circle*

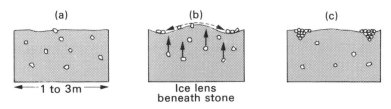

(a)   (b)   (c)

◄—1 to 3m—►   Ice lens
                                beneath stone

stones. On sloping ground the same process operates, but the circles are elongated, until on very steep slopes they become straight lines perpendicular to the contours, called stripes.

Ice wedges

When the ground is frozen hard, after an initial expansion it will begin to contract, and in so doing cracks will form. If there is permafrost below an active layer, during surface thaw water may run down into the cracks in the permafrost and freeze, constituting lines of future weakness. During the following winter the cracks will reappear, to be filled with more water in the next thaw. In this way wedges of ice will slowly form (Fig. 5.4). When the ice is only a few millimetres thick it is called a vein. As the veins become thicker they usually taper downward and are called wedges. If the permafrost melts, the gaps left by the ice wedges will be filled with material from above. This might be of a fine-grained nature moved by wind or rain, or stones and gravel transported by meltwater. Differences in soil or subsoil, as a result of periglacial processes operating during the Pleistocene and widespread in the now temperate lands of the northern hemisphere, are often clearly visible on air photos because they result in differences in vegetation type, or rate of growth, or colour. Surface cracks — and therefore ice wedges — typically take the form of rough hexagons. Wedges have been identified up to 10 m across at the top penetrating 10 m into the ground. The polygons may be up to 30 m in diameter.

Photos 5.3 and 5.4, taken recently in Arctic Canada, show respectively stone circles and a hexagonal pattern resulting from the formation of ice wedges. A surface marked by these kinds of features formed as a result of cryoturbation is generally referred to as patterned ground.

Air photos are one source of evidence that similar conditions once existed in Britain. Photos 5.5a and 5.5b are of natural vegetation taken on infra-red film which shows green growth as a light tone (the reverse of normal panchromatic film). The darker areas are of heather growing in deep sandy soil. The light areas are grass; here chalky rubble is much nearer the surface, giving an alkaline soil in which the heather cannot grow. The polygons disclosed in Photo 5.5a (mostly rough hexagons the larger of which are about 10 m in diameter) are the results of sand infilling

Fig. 5.4 *Evolution of an ice wedge (see text). The material displaced by the ice is forced upward making the ground above uneven*

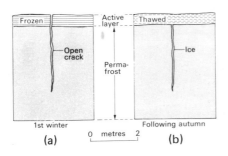
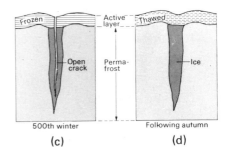

Photo 5.3 *Example of stone circles in limestone from Arctic Canada* P. James

Photo 5.4 *Ice wedge hexagons in Arctic Canada* P. James

Photo 5.5 *Patterned ground (see text) resulting from previous periglacial conditions:*
(above) *on Thetford Heath, Elveden, Norfolk,*
(below) *near Grimes Graves, Norfolk*

Crown copyright. Ministry of Agriculture, Fisheries, and Food

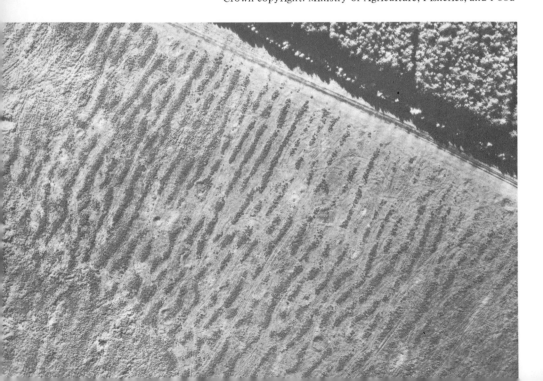

cracks in calcareous glacial drift after ice wedges melted towards the end of the Pleistocene. The stripes shown by the heather in Photo 5.5b are about 3–4 m wide and of striking regularity. They are almost certainly the result of periglacial processes, but a satisfactory explanation of their formation has yet to be suggested.

Pingos

Pingo is an Eskimo word for scattered, isolated dome-shaped hills. They occur in areas of discontinuous permafrost and vary from 1 m to over 60 m in height, with a diameter up to 300 m. The tops of some are rounded; some have a crater-like depression at the top.

The formation of a pingo cannot be explained in terms of frost heave, because it is typically composed of a high proportion of sand, which is not susceptible to heave. Indeed there is no generally accepted theory for their formation. The method illustrated in Fig. 5.5 provides a possible origin. One alternative theory is that hydrostatic pressure, building up beneath a surface layer of permafrost, forces water to the surface at a weak point, and the pingo is formed by deposition of material carried up from below. Both theories lack sufficient evidence for any firm conclusion to be made.

It has been suggested that some circular or semi-circular ramparts with diameters up to 120 m found in Wales, Belgium, and West Germany may be fossil pingos. (They were not made by man.) The possibility that similar features may have a periglacial origin should be considered when met with in the field.

Fig. 5.5 *Pingo formation in the Mackenzie delta area.*
(a) *Broad shallow lake over frost-free ground*
(b) *Accumulation of sediment in lake causes lake ice to freeze to bottom in winter. Edge of permafrost advancing, causing expulsion of pore water in the saturated sands and giving upward pressure beneath lake centre*
(c) *Further advance of permafrost; up-doming of lake ice and sediment*
(d) *Permafrost continuous; up-doming has ceased, and ice-cored pingo remains*

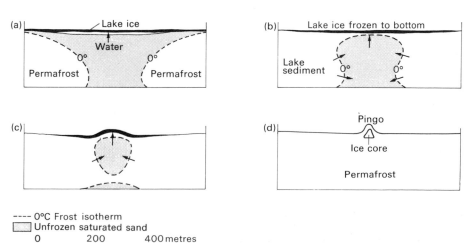

Thermokarst

This is a term used to describe the features of a previous area of permafrost from which the ground ice has melted. The removal of subsurface ice leaves dry valleys, pits, and funnel-shaped sinks resembling the classical karst landscape. Possible fossil thermokarst areas have been described in the Paris Basin, and it is as well to bear in mind the possibility of finding such features in Britain.

Slopes in periglacial conditions

One of the differences between slopes developed in temperate and in periglacial conditions is the importance of solifluction, especially in association with permafrost. Annual movements of solifluction material provide a far more rapid mechanism for general slope modification than any process in temperate areas (other than the catastrophic occurrences of landslide, slip, and flow which are confined to extremes of slope and weather). Slopes may thus become considerably altered over relatively short periods of time. The degree of solifluction depends in part on expansion and contraction, and the formation of ice crystals in the soil as a result of freeze/thaw, but more importantly on the presence of moisture. There is evidence that moisture content rather than slope angle determines the rate of flow. Although solifluction may be generally anticipated on steep slopes of between 15° and 35° it

Photo 5.6 *Section through the subsoil in Lambley valley, Nottinghamshire. Angularity of the stones is evidence of frost shattering. The unsorted nature of the material is proof of solifluction and the periglacial processes operating in the Pleistocene*

can also be effective on slopes with angles as low as 2° or 3° provided sufficient water is present, especially if the material concerned has a fine particle size. For this reason considerable surface movement can be expected on quite gentle slopes in periglacial areas where the surface is kept continually wet by rain, or melting snow or ice, or by other means during the period of thaw. These are important considerations when attempting to interpret present landforms in places where periglacial conditions are known to have previously existed.

As we have seen (p. 61) the detailed study of slope processes is of recent origin, and we cannot be sure of the extent to which our present landscape is in part a relict of the past. What is beyond doubt is that in evaluating present process we must always keep before us very clearly the possibility that the landforms we are studying may in their main outline be the result of processes not operating today. Photo 5.6 shows a section through the subsoil in Lambley valley in Nottinghamshire, an area not occupied by the final ice sheet, but which must have experienced severe periglacial conditions. The material exposed is derived from Keuper Marl. Most of it is clayey marl, while the stone fragments come from the associated sandstone skerry bands. The angularity of the stones show they must have been frost-shattered. The unsorted nature of the deposit is evidence that it was not waterborne, but could only have been the result of solifluction. The whole valley floor is covered with it to some depth. (Photo 2.3a gives a general view of the valley.) There is no doubt that any interpretation of landforms in this valley, and the processes now operating, must take into account the periglacial processes of the past.

Glaciation

There have been a number of major ice ages in the past dating back to the Pre-Cambrian, and even more theories for their cause. Continental drift has become a general favourite to account for the distribution of glacial tillites and striated rocks of Permo-Carboniferous age in Africa, and the presumption seems reasonable in this case. However, we have to rule out continental drift as the only cause of the most recent ice age, and until there is more knowledge of the reasons for the fluctuations in temperature which gave rise alternately to cold (glacial) and warm (interglacial) spells, all we can do is to confess we do not know. Neither is there any general agreement about the precise date of the onset of the recent ice age. Early estimates were that it began very roughly one million years ago. In fact, there is evidence that the ice did not begin to accumulate everywhere at the same date. (In the special case of the Antarctic, glaciation probably may have begun as long as 11 million years ago.) The time of the appearance of ice sheets may have varied considerably in different parts of the northern hemisphere. Today it is believed the onset of the northern ice began certainly more than 2 million years B.P. (Before Present).

Previously it has generally been assumed that the fall in temperature associated with the onset of an ice age would be gradual. Recent research has shown that this assumption is probably not valid, and there is growing evidence that a fall of temperature sufficient to initiate glaciation in temperate latitudes could occur over as short a period as 100 years. If this conclusion is confirmed there would seem little point in attempts to predict future glaciations from the analysis of very long-term trends.

It is probable that only marginal changes are required to initiate a period of glaciation similar to those of the recent past. The latest cycle of cooler climate dates from between the end of the fifteenth and the end of the nineteenth centuries. During this period most Alpine glaciers were at their maximum extent in historic time. There is now evidence to indicate that conditions about the middle of the cycle were very close indeed to those required for the initiation of a major glaciation, and that we find ourselves, almost by chance, in what may well still prove to be only an interglacial.

Glaciers

When snow first falls on a cold surface it settles as powdery flakes. As accumulation proceeds the snow underneath is compressed by the layers above and becomes consolidated (but not yet ice). This snow is called firn. (It is generally accepted that, until it becomes ice, snow more than one year old is firn.) Slight snow-melt during the daytime accelerates the process of consolidation. The density of newly fallen snow is about 0·07; the density of firn (or *névé* in French) is about 0·5. As the firn becomes thicker the lower layers are compressed and the air content reduced until eventually a stage is reached when it is impermeable to water. At this stage it may be considered ice. The time taken for this to happen varies according to the location of the glacier. A minimum of 25 years has been found necessary in the temperate latitudes of Switzerland, whereas over 150 years are required in the polar regions of Greenland. Snow sheets tend to be self-perpetuating, because once unmelted clean snow lies on the ground a high proportion of solar radiation is reflected directly back into space and the warming effect of the sun is reduced. Snow is nevertheless a good insulator, due to the quantity of air trapped between the ice crystals. It helps to protect the ground from cold and may itself be warmed by geothermal heat. (The insulating effect of snow makes possible the sowing of winter wheat in Canada.)

Snow or firn may be subject to slow creep, or to catastrophic movement as an avalanche. Snow can hold up to 75 per cent of its own weight of water and should the interstitial air be displaced by rain, or snow-melt due to a rise in temperature, it is able to move as a viscous fluid. The resulting friction releases heat, which melts the points of the ice crystals, allowing movement between the grains. If the slope is steep enough the additional weight of water and the reduced cohesion in the snow can result in an avalanche.

For a snow patch to become the source of a valley glacier the average annual accumulation of snow must exceed the loss. The lower layers of firn slowly thicken and become ice. If the process continues, at a critical thickness some of the ice will eventually begin to move downhill in the form of a glacier. This source area, from which the glacier continues to be fed, is known as the accumulation zone. Sometimes the descent from the accumulation zone may suddenly steepen, in which case the glacier passes down an ice-fall, an area of rapid movement with many crevasses and much broken ice. After the ice leaves the accumulation zone, at some point it passes what is called the equilibrium line (which may vary from year to year) where net accumulation of ice and snow is balanced by net loss. The loss of ice from all causes by a glacier is known as ablation, and is caused by melting, evaporation, wind erosion, avalanches, and calving (the floating away of large blocks of ice in the form of icebergs where the snout of a glacier reaches water). The balance between accumulation and ablation is known as the glacier's budget. The larger the gross budget, that is the higher the rate of accumulation and ablation, the faster will be the speed of the ice. Fast-flowing glaciers therefore are those of the more temperate latitudes where there is ample precipitation and temperatures rise enough in summer to give high rates of ablation, for example on the western side of the Southern Alps of New Zealand, in Switzerland, Iceland, and southern Norway. These may be compared with some of the Antarctic glaciers, where precipitation is low, and as the temperature rarely rises to freezing-point the only effective form of ablation is through calving.

If, over a period, net accumulation consistently exceeds ablation, the length of the glacier will be extended, that is it will thicken and grow. This may happen through an increase in precipitation, or a reduction in the rate of ablation due to lower temperatures, or both. If these processes are reversed the length of the glacier will be reduced. It is misleading to say the glacier will retreat. Neither a glacier nor an ice sheet can move backwards. What happens is that the downwards flow of ice is not sufficient to balance the ablation in the lower part of the glacier. The ice therefore shrinks steadily and finally disappears, leaving behind the mass of debris called the terminal moraine, while the glacier's snout occurs at a point where the supply of ice is just adequate to maintain it. Therefore although the ice cannot move back upvalley the lower end or snout of the glacier may rightly be said to do so.

Glaciers may be divided into three main categories:

1. temperate glaciers, like those of the Alps and southern Scandinavia;
2. cold or polar glaciers;
3. a few which possess characteristics of (1) and (2).

The differences are important. In temperate glaciers meltwater exists permanently or seasonally beneath the ice, so that the glacier can flow more easily over the wet rock. Futhermore the flow of subglacial meltwater, sometimes under considerable hydrostatic pressure, can cause substantial erosion. A cold glacier has no meltwater between the ice and the rock on which it rests. Even though summer meltwater streams are found at the edge of the south Greenland ice sheet, they flow within the ice and not below it. The cold glacier is therefore frozen to the rock on which it rests. This affects the movement of the ice and may considerably reduce (or perhaps eliminate) erosion.

Most glaciers today are in a state of recession, having been at their historic maximum (mentioned above) in the early eighteenth century. A few glaciers have recently advanced. One of the most spectacular was the Behring glacier which advanced up to 1200 m between 1963 and 1967 on a front 42 km wide. Surges like this (and those of smaller magnitude) must initially be due to a local climatic fluctuation resulting in a positive increase in the glacier's budget, although it may be only after many years of accumulation that the glacier reacts. The exact conditions which bring the glacier to a critical point at which a surge occurs are not fully understood. It has been suggested that subglacial temperature and pressure 'may be important factors. As the ice thickens so pressure at the base of the glacier increases. When the pressure becomes great enough, increased melting of ice in contact with the valley floor will result (see page 95). If this happens over a sufficiently large part of the glacier, the reduction in friction could cause a sudden increase in velocity and hence a surge. Because of the increase in velocity the thickness of the ice, and consequently the pressure, is reduced. The basal ice then begins to refreeze, and a new state of approximate equilibrium returns. It has also been suggested that some form of trigger mechanism is involved at the end of a period of stability. Earthquakes are quoted as a possibility, and it may be significant that the main surges this century have occurred in areas of seismic activity.

A different kind of surge has also been observed to occur in some glaciers. As described above the ice first deepens in the accumulation zone, but this time suddenly discharges in the form of a kinematic wave. That is, the excess accumulation of ice passes in a single wave form down the glacier. The wave compresses the ice in front of it, causing the depth of ice in the glacier to increase sharply. As the wave passes, the ice behind is stretched or tensioned, giving rise to a band of crevasses. The speed of the wave, initially faster than the movement of the glacier, decreases as it travels downwards. At the same time the height of the wave increases so that its effect is greatest near the snout. After the wave has passed the main glacier returns to its previous level, but the added ice pushes the snout to a new point further down the valley. Whether this location remains permanent depends on whether the glacier's budget remains in equilibrium.

Ice movement and glacial erosion

We have seen that because of the high rate of accumulation and ablation of temperate glaciers (i.e. because they have a large gross budget), the speed of ice movement downhill is generally much faster than that in cold glaciers. Considerable variations may also exist between glaciers, between different periods of time for the same glacier, and at different points on the same glacier's long profile. Sudden surges can occur, but most glaciers move at an average speed of less than 1 m a day.

The velocity of the ice in a valley glacier, like the velocity of water in a stream, varies over its cross-section. (Measurements are made by boring holes through the ice, inserting tubes, and recording their subsequent deformation.) The variation in speed is caused by friction with bed and sides. The highest velocities are found at the surface in the centre. Under pressure ice behaves like a plastic, with the ability to flow. Thus friction at the rock face can be transferred through the deformation of unbroken ice. Where tension is great the ice may be broken by the formation of crevasses. Studies have shown that ice velocities decrease from the centre of the glacier, at first slowly and then more rapidly as the distance from the rock face is reduced. In very fast-moving glaciers, most of the ice may move at roughly the same speed with the reduction in velocity confined to a zone close to the rock contact face. This is termed *Blockschollen* flow (Fig. 5.6), and normally results in greatly

Fig. 5.6 *Distribution of vertical* (a) *and transverse* (b) *velocity in a typical temperate glacier of normal and Blockschollen types of flow*

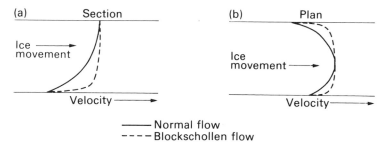

increased erosion. The tension created along the edge of the ice is immense, and if the glacier bed is irregular it breaks the surface at the sides into blocks, or seracs.

It is useful to consider the movement of ice within the glacier in association with the long profile of its valley. Typically the long profile of a glaciated valley descends in a series of steps consisting of hollows, sometimes infilled with sediments or drift, and separated by rock bars running across the valley floor. Occasionally the hollow is occupied by a lake, but more often the rock bars have been cut through by a stream which drains the whole system, probably initiated by subglacial melt-water. The bars are formed from bedrock and often exhibit smoothed surfaces on the upvalley side, frequently striated by moving ice, with plucked and roughened lee sides to give a *roches moutonnées* effect.

Most temperate glaciers have occupied previously stream-eroded valleys, and in the early stages of growth the movement of the ice must have been largely controlled by the profile of the valley. J.F. Nye has suggested that where there is a reduction in gradient in a valley floor the ice in the glacier will tend to decelerate and become thicker. This he calls compressing flow. In places where the valley floor steepens the ice will accelerate and become thinner, called by him extending flow. Let us consider the simple hypothetical long section of part of a glacier shown in Fig. 5.7. As the ice moves from A to B, gradient is reduced, velocity may decrease a little, but the glacier becomes thicker. As the valley slope increases again in C the ice will tend to accelerate and the glacier become thinner.

There are a number of things which affect the erosional capacity of a glacier. The two most important are the longitudinal velocity of ice movement and its

Fig. 5.7 *Extending and compressing flow of ice within a glacier (see text)*

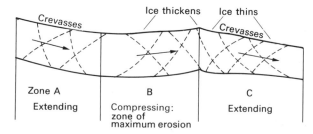

thickness, the latter controlling the pressure exerted on the rock bed. Erosion depends primarily on the balance between these two factors. Sections of the valley where the ice is thicker will tend to be eroded more effectively. More erosion will therefore take place in zone B in Fig. 5.7 than in A or C. In this way an existing irregularity in a valley floor (for example a pre-existing knick point) may be increased by the effect of glaciation. This process provides a situation in which positive feedback is taking place. The feedback is positive because increased erosion in area B will lead (up to a critical limit) to still thicker ice and therefore to still greater erosion. In other words the process is cumulative, and the existing hollow is progressively enlarged (cf. negative feedback, p. 55).

It will be seen that Fig. 5.7 shows a number of slip planes passing through the ice. These are possible lines of weakness indicating potential sheer surfaces associated with the two different types of flow. When flow is compressing the main slip planes extend tangentially upwards from the bed in a downvalley direction. Movement along these planes could therefore carry rocks up from the valley floor into the ice, arming the glacier, and making it a much more effective agent of erosion. In areas where flow is extending (accelerating) the main slip planes will tend to develop from the surface curving downvalley to meet the bed tangentially.

Many factors may contribute towards the formation of hollows and rock bars. The addition of ice to a main glacier at its junction with a tributary, or a reduction in the width of the valley, are examples of conditions which may lead to changes in the velocity and thickness of the ice. Lithological differences could be important in the formation of a hollow when ice moves across alternately weak and resistant rocks. Once these features are created, if not removed by other processes during interglacials, they would be accentuated by succeeding glaciations. But as the height of the bar increases it presents more and more resistance to the ice and greater erosion takes place, placing a limit to the ultimate height of the bar.

Temperate glaciers discharge meltwater, often through ice tunnels formed between rock and ice, rather like water in a pipe. Because the tunnels are enclosed, great hydrostatic pressure can be created with the result that the subglacial streams may flow uphill as well as down. (Their relict channels can sometimes be identified

Fig. 5.8 *Valley floor shattering under periglacial conditions and subsequent glacial erosion*

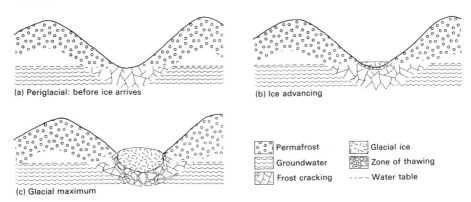

(a) Periglacial: before ice arrives

(b) Ice advancing

(c) Glacial maximum

Permafrost   Glacial ice

Groundwater   Zone of thawing

Frost cracking   ---- Water table

by this means.) The erosive effect of fast-moving water under pressure and charged with debris can be considerable.

A very interesting theory to help account for the enormous amount of glacial erosion known to have taken place during the Pleistocene is illustrated in Fig. 5.8. The theory assumes a period of periglacial conditions before the arrival of the ice of sufficient duration for permafrost to form in depth. The bottom of the valley is considered to be wetter than the valley slopes, partly because of the relative height of the water table, and partly because water would tend to run down the hillside to the valley floor. Water and frost together could cause deep shattering in the rock. As the valley became ice-filled the floor would become insulated from the cold and tend to thaw (through heat conducted upwards from below), leaving the already broken rock to be rapidly eroded first by the ice and later by meltwater as the glacial period came to an end.

Another theory is that of pressure release. As the glacier carries away quantities of material from the surface, rock of a density of 2·5 or more is replaced by ice with a density of about 0·9. This release of pressure on the rocks beneath can cause cracks to appear parallel to the surface. (This phenomenon is called dilation jointing and is also observed in opencast mining.) These are planes through which water can penetrate and subsequent freeze/thaw action provides fresh quantities of shattered rock for the ice to remove. On deglaciation any broken rock remaining would be exposed to reduction by freeze/thaw action and, with great quantities of meltwater to carry the fragments away, erosion could be rapid.

The precise mechanism of glacial erosion is still not known, largely because of the extreme difficulties involved in tunnelling into the base of glaciers to find out. However, evidence collected over the last twenty years makes a number of theories possible. Under pressure ice behaves like a plastic, that is, it will resist stress up to a critical point and then deform. An analogy is that it behaves like mild steel heated to a temperature at which it becomes soft. The amount of pressure required to be applied when the resistance of the ice just begins to fail is known as the sheer stress. It has been found, largely through laboratory experiments, that the lower the temperature the greater the stress required before the ice will behave as a plastic. It has also been found that the method of formation of the ice is important. For example, at the same temperature lake ice derived directly from water is much stronger than glacier ice formed from the accumulation of snow.

Like all solids the melting-point of ice is dependent on both temperature and pressure. A simple experiment to demonstrate this is to suspend weights from a block of ice by means of a wire loop. When the weights are sufficiently heavy the ice will melt beneath the pressure of the wire and allow it to pass slowly through the block. The ice reforms as the wire passes and the block remains unchanged. The same happens when a small metal cube is placed beneath the wire. It will be drawn through the block without damaging the ice.

A temperate glacier moves partly through internal flow within the ice and partly through slip over the valley sides and floor. From the few observations that have been made by tunnelling underneath temperate glaciers, it appears that two main processes are involved. Firstly there is the ability of the ice under pressure to flow round small obstructions. For example, imagine a boulder protruding from the glacier's bed. The ice moving past is forced up and to either side; because it behaves like a plastic it will reform on the downvalley side of the boulder. Secondly temperature and pressure at the base of the glacier may be such that some melting occurs. As pressure increases so the temperature at which ice will melt is lowered, and is commonly called 'the pressure melting-point'. As the ice reaches the boulder pressure increases and the ice may reach pressure melting-point, with the result that a film of water between ice and rock lubricates the passage of the ice. (The same kind of process happens in ice skating.) As the ice moves past the boulder it refreezes. This process has been called regelation slip. There is now evidence concerning temperate glaciers to confirm that the nature of the lowest layer of ice in contact with the rock is consistent with melting and regelation under conditions of changing pressure. Some at least of the meltwater associated with temperate glaciers may originate in this manner and it is certainly a very important factor in assisting movement, and in the capacity for erosion. Regelation is probably important in ice plucking – the ability of moving ice to freeze boulders and rock fragments within itself. And there is no doubt that armed in this way the base of a glacier with the great weight of ice above it is a formidable tool for excavation.

Far more is known about temperate than cold glaciers. Cold glaciers, it will be remembered, are those frozen to the rock on which they lie. If meltwater streams exist for short periods these flow within or on the ice and not on bedrock. Because they are frozen to their bed, movement seems to be entirely through plastic flow internally within the glacier. It is generally, though not universally, agreed that cold glaciers cannot cause any appreciable amount of erosion. However, for lack of evidence these conclusions must remain very tentative.

Glacial deposition

Glacial deposits, usually termed glacial drift, are important because they cover a large area of the north temperate landscape. They frequently provide a sizeable proportion of the load transported by streams and may exert considerable influence on agriculture. They may be divided into two categories, determined by their origin:

1. **Fluvio-glacial deposits**. These are laid down by meltwater and are sorted. They may also be stratified. They comprise features like eskers, and deposits such as outwash sands and gravels.
2. **Glacial deposits**. These are deposited directly by the ice and are therefore unsorted, e.g. moraines.

(Boulder clay is a term sometimes used to define unsorted glacial drift. It is unfortunate that the Geological Survey has perpetuated this name as boulders and clay are rarely found together. The term glacial till is much to be preferred in its place.)

In Europe and North America glaciation probably started mainly in upland areas, but snowfall remained sufficiently high for ice sheets to develop and flow down over large areas of lowland. Although there are depositional features in the uplands the most obvious glacial features there are erosional. In the lowlands the result of glaciation is mainly deposition, sometimes accompanied by changes in the pattern of drainage.

All Britain was covered by the Pleistocene ice down to a line running (very approximately) from the north Devon coast in the west through Oxford to Essex in the east. North of this line fieldwork can provide much useful information in lowland areas by the examination of drift. The interpretation of complex deposits can be very difficult, but much good work can be done on the simple analysis of field observations, for example, the orientation and dip of the long axes of pebbles, or pebble roundness coefficients, or on an observed sequence through a terminal moraine, and in many other ways. (One technique is exemplified in *Science in Geography 4: Data Use and Interpretation*, p. 43. Other techniques described elsewhere in the book can be readily adapted for use.)

Features of glacial deposition are generally well described in current literature and are not dealt with here. But the origin of one particular landform, the drumlin, is still unclear. All are agreed that drumlins are formed below an ice sheet and most are agreed that their orientation is in accordance with the direction of ice movement. Thereafter agreement ends. The drumlin has a characteristic shape (Fig. 5.9). They may attain a height of 60 m, with normally a length to breadth ratio of between 2·5:1 and 4:1. They may consist of almost any material, stratified or unstratified, or both, and some are of solid rock. They may exist in isolation, but they are usually found in clusters or 'swarms'. They abound in some previously glaciated areas and are completely absent in others. Drumlins of solid rock are found existing together with others of sand and clay.

A drumlin of solid rock is obviously an erosional feature, though exactly how this occurs is unclear. Normally rock eroded by ice displays a smoothly curved surface on the side from which the ice advanced, with a plucked and roughened surface on the other, like a typical *roche moutonnée*. The drumlin has a streamlined tail on the downglacier side and must therefore have been formed partly in a different way.

It has been suggested that drumlins might be formed during a period of fluctuations in the extent of an ice sheet as a product of the ice's readvancing over its own deposits, which might be unsorted moraine or stratified fluvio-glacial material. This seems satisfactory, but does not explain the formation of the steep stoss side by the advancing ice. If anything the reverse might be expected.

Another suggestion is that frozen till or rock forms a nucleus and the drumlin is built up around it layer by layer as more material is frozen on to it from the base of the ice sheet. There has also been a theory that as an ice sheet passes over a boulder

Fig. 5.9 *A typical drumlin*

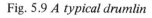
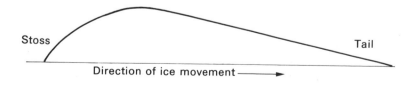

a drumlin-shaped hollow is created in the ice and till forced up to fill it. Unfortunately, for this theory to hold, the ice would have to be very slow-moving or stagnant, and drumlins have been described when this condition can be shown not to be met. In fact, there is evidence that drumlins tend to form when pressure is greatest, caused by a local thickening of the ice.

Perhaps the only thing that can be said with certainty is that if the orientation of the drumlin correlates significantly with the orientation of the long axes of stones within it, then both must have been laid down by the same moving ice. Statistical testing between groups of characteristics such as length:breadth ratios, stone roundness and orientation, height and composition, can provide a fruitful object of field studies. Most investigations require work in the field, but some can be done from a 1:25 000 O.S. map. Much is still to be learned. Almost all that can be said of the evolution of drumlins is that we know practically nothing about it.

Consolidation

1. What is meant by a periglacial climate? Where do these exist today? Where would they have been during glacial periods of the Pleistocene?
2. What processes are dominant in periglacial conditions? How do they affect erosion and transport?
3. Describe, and suggest explanations for, landforms which result from periglacial processes.
4. How do glaciers move? How does this knowledge help to explain the long profile of a typical glaciated valley?
5. What is meant by a glacier's budget and why is it important for the behaviour of the glacier? What happens if the budget changes?
6. Consider the importance of glacial meltwater, and the landscape features which result.
7. What is known of the process of glacial erosion?

Chapter 6

Process in warm arid and semi–arid areas

Warm semi-arid climates

Attempts have been made in the past to associate different kinds of climate with typical landform features. In particular inselbergs and pediments have caught the imagination of geomorphologists in a way few other features have done. Explanations for their form and maintenance have frequently been sought in the nature of the climatic environment in which they have been studied. In Germany the word *Klimamorphologie* was coined for this kind of investigation. It is now recognized that research along these lines is unlikely to prove very fruitful, if only because of the comparatively recent and widespread changes in climate. For example, there is much evidence that there have been major pluvial periods during the Pleistocene in what are now semi-arid parts of the world. It is therefore not possible to assume that present processes were active throughout the past. Many different types of landform are also present under climatic conditions which are relatively homogeneous over wide areas. Many of the kind of landforms common in semi-arid areas are also found under temperate humid conditions. For these reasons, present research in this field today, as in other landform studies, tends to concentrate on present processes rather than on evolution.

For geomorphic purposes a semi-arid climate may be defined either as one in which there are one or two distinct dry seasons during the year when vegetation dies back leaving patches of bare earth, or as one in which precipitation is insufficient to maintain a complete ground cover by vegetation at any time. The importance of this is that the binding effect of roots and the protection from the impact of raindrops afforded by grasses and other close vegetation cover is absent on part of the soil at some time every year. Rain and wind are thus able to attack the soil directly and erosion is correspondingly more effective.

The rainfall regime of the tropical semi-arid lands is typically concentrated into heavy convectional storms. The effect is quickly to saturate the ground and produce a situation in which, during the rainy season, there is frequent surface runoff, sometimes accompanied by flash floods. (See p. 36 and Fig. 3.6.) The processes most active are thus removal by wind, strong surface runoff, the direct attack by rain on soil, and chemical decomposition accelerated by heat.

Fig. 6.1 *Two common types of inselberg.*
(a) *Composed of bare rock and making an abrupt angle with the plain below. The pediment and hillslope are formed of homogeneous rock and are erosional features. The bahada, forming typically on the pediment, is a result of deposition*
(b) *A free face with a rectilinear scree slope below it. The scree frequently makes a definite angle with the plain, as a waning footslope is often absent. If the free face is formed of a resistant caprock a small inselberg may be termed a butte, or, if of considerable area, a mesa*

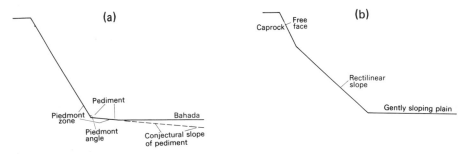

Inselbergs

Inselbergs (meaning island mountains) are steep isolated hills usually standing in a gently sloping plain. They may simply consist of piles of boulders, in which case they are known as kopjes. They may also appear as a steep-sided rock or wall or dome with no regolith, or as a free face with a rectilinear slope of shattered rock below. Where the regolith is absent there may be found a gently sloping pediment of bare rock at the base of the main slope. Lower down the pediment is typically covered by a depositional slope called a bajada, or bahada (Fig. 6.1).

Some inselbergs are most probably caused by a difference in rock type. In Africa they are characteristically — though not invariably — composed of a granitic type of rock. If this rock is surrounded by material that is more easily decomposed an inselberg will eventually result. The difficulty in proving this is the depth of the regolith on the surrounding plain. In much of Africa the land is very ancient and weathering has been continuing for very long periods indeed. Temperatures are also high in the intertropical zone and other than in some very arid areas the bedrock has often been decomposed to great depths. (In equatorial Singapore island, decomposition exceeds 100 m or more in many places.) It is thus impracticable in many cases to establish the nature of the original surrounding rock.

One of the problems concerning inselbergs is to account for the abrupt change of angle between the hillslope and the plain. Fig. 6.1a represents an inselberg developed in homogeneous rock in which the piedmont angle is not structurally controlled (i.e. it is not the result of a fault or shatter zone). We are not here concerned with the way in which the slope evolved to its present state — that must be largely speculative — but we are interested in the processes at present at work upon it which result in the maintenance of the abrupt angle at its base. This change of angle, cut in homogeneous rock, is not usually found associated with a temperate climate, except in some sea cliffs where at high water the sea provides unimpeded removal at the base of the cliff. Under temperate conditions, in the absence of removal (e.g. by a stream) deposition tends to take place at the bottom of the slope followed by the formation of soil fixed by vegetation, which covers the bare rock and slowly extends upwards. The maintenance of an abrupt change of angle seems to be associated with semi-arid areas (as defined above) and it is worth considering in some detail the active processes involved.

The hillslope, piedmont angle, and pediment are erosional features; the bahada is depositional. The hillslope is often approximately rectilinear. For this profile to be maintained removal from it must everywhere be unimpeded. In hot areas, where water is present either in the form of rain or dew, heat intensifies the chemical reactions which take place over the whole surface of the slope. In a granitic type of rock the hard crystalline particles are loosened through the chemical decomposition of the surrounding material, a process which may be assisted by the often large diurnal temperature range. On the hillslope loosened particles, depending on their size, may be transported directly away by the wind in the form of dust (anyone who has experienced a tropical duststorm will confirm the large amounts of material that can be moved in this way) or roll or be washed down to the base as grains of sand. From here they may possibly be removed by saltation (page 105) or possibly by the vigorous surface wash typical of the rainy season. In other words the weathered products from the hillslope are rapidly removed from the whole rock surface and consequently the slope retreats parallel to itself and retains the same profile.

Weathering also takes place on the pediment. Here removal is not so efficient as on the hillslope. Although dust and some sand may be transported by the wind, because of the gentle angle larger grains tend to remain longer in position before removal by surface wash, and erosion is thus slowed down. There is also an argument that reduced erosion on the pediment is partly because surface wash on the steep hillslope is in a condition of rapid flow with $F > 1$ (page 52), whereas at the base flow will return to normal with $F < 1$. It seems reasonable that such a change in the erosive capacity of wash would help to maintain the piedmont angle. The pediment may be regarded as an erosional slope of transport on which material from the hillslope above is conveyed to the depositional slope of the bahada below. The angle of slope of the pediment is probably related to the rate of erosion of the hillslope above, while that of the bahada to the size of the material of which it is composed and the quantity and vigour of the surface wash. In contrast to the transportation slopes of temperate humid regions the erosion of both the hillslope and the pediment are controlled by weathering. Most pediments are slightly concave. This reduction in gradient away from the hillslope base may be because surface wash increases as the catchment area for rainfall grows, and the larger volume of water is able to carry its load down a gentler slope. It is also probable that as distance from the hillslope increases the size of particles will be reduced and therefore become more easily transportable. The bahada starts at the point when reduction in gradient is just sufficient for deposition to begin.

One theory to account for the type of slope profile described above is that the hillslope, pediment, and bahada have achieved an equilibrium, and as the hillslope retreats so the pediment grows and with it the bahada, fed by the deposition of

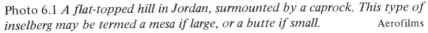

Photo 6.1 *A flat-topped hill in Jordan, surmounted by a caprock. This type of inselberg may be termed a mesa if large, or a butte if small.*   Aerofilms

material weathered from above. In this way the inselberg will eventually be consumed, leaving a gently sloping plain behind. There are many examples where this kind of progression appears to be in its final stages. But, for the mechanism to work, it is necessary that rapid removal takes place from the hillslope and pediment. This can only occur in areas where grasses die back during part of the year to allow wind to remove the finer particles of weathered rock and inhibit the development of soil in the piedmont zone.

The profile shown in Fig. 6.1b is another kind of slope often found in association with an inselberg, and a form that is quite common in semi-arid parts of the world. It develops typically after a long period of erosion when remnants of a hard resistant horizontally bedded stratum form the caprock to isolated hills. The resistant caprock makes a free face with a scree slope below it, generally rectilinear in profile, of material derived from the free face (Photo 6.1). The slope angle will depend principally on the size of the rock fragments composing it. Slopes with a similar profile, although of quite different origin, are common in Britain where they often form the sides of glaciated valleys, with the gentle slope of the valley floor on which the scree rests the result of fluvio-glacial deposition. They are also found along the Millstone Grit edges in the Peak District. Here the massive gritstone overlying shale creates a free face or edge, with the scree slope below caused through frost shattering, mostly under periglacial or near-periglacial conditions. Whether in temperate Britain, or in a savanna-type climate, while there is sufficient material weathered from the free face for the scree actively to grow, the break of slope at the base will be maintained. But in a temperate area once this process ceases soil begins to form and, unless there is removal by a basal stream, a waning slope develops. Under warm semi-arid conditions this does not seem to happen so effectively, and the angle between scree and plain tends to remain distinct, even when the slopes are sufficiently stable to have become fully or partly vegetated. It is possible that two factors contribute to this. Firstly, the material of the scree slope, being mainly from the caprock, is different from that of the plain (a factor which may affect slope angle in temperate areas too). Secondly, and probably more significant, is the removal (by the process described above) of fine particles from the base of the scree by wind in areas which have a distinct dry season.

Weathering in arid areas

Blown sand will tend slightly to undercut rock outcrops very close to the ground, but the strangely shaped desert *Zeugen* (things) are not the product of sand blasting. It has been demonstrated that sand (as opposed to dust) is rarely carried by the wind more than 2 m above ground level, and movement is mostly confined to less than 1 m. The wind storms of desert areas that produce the much photographed towering clouds of dry particles hundreds of metres high, darkening the sky so that artificial light is required to read at noon, and generally termed sandstorms, are in fact clouds of fine impalpable dust, quite incapable of eroding anything. The Touareg on his camel is out of reach of the sand, other than under exceptional conditions; his face cloth is protection against dust. Removal of fine particles by wind is a very important aspect of denudation in arid areas. It has been reported that in 1863 four million m^3 of dust fell on the Canary Islands, borne by an easterly wind from the Sahara across between 100 km and 400 km of ocean.

It may seem a contradiction to state that little erosion can take place in deserts without the presence of moisture, but such is the case. The climates of hot deserts are marked by aridity and a very large diurnal temperature range. This range is found even in coastal deserts such as the Namib, but is of course much greater in continental interiors like most of the Sahara. The air contains insufficient moisture for cloud to form and, unless there are strong winds to raise dust, remains very clear. Consequently, during the day, temperature on the surface can be very high indeed. At night radiation is assisted by the clear air, and temperature falls very rapidly. Although frost may be rare the air near the ground becomes cold enough for dew to form and this provides the moisture necessary for the chemical reactions associated with the decomposition of rock. It should be noted there is no evidence that temperature change alone, great though the diurnal range is, causes mechanical weathering. Expansion and contraction may however cause minute hairline cracks to appear between the granules composing the rock and thus allow the penetration of moisture. The result is chemical decomposition accelerated by the desert heat and loosening of the rock grains. This is called granular disintegration and is typical of granitic type rocks, though it occurs in other rocks including sandstones.

Alternatively exfoliation may be initiated in homogeneous fine-grained rocks not subject to granular disintegration. Moisture enters the rock through minute cracks and if it penetrates to a depth of a few centimetres it will be retained longer than moisture near the surface. In this way a decomposed layer is initiated below the surface to produce typical 'onion' weathering. The solid products of decomposition are sometimes bulkier than the original material, creating pressures which result in surface layers of rock becoming detached. (A somewhat similar effect is produced by pressure release (page 95) and the two should not be confused.) Ferrosilicic salts in solution (products of decomposition) may be drawn to the surface by capillarity and the moisture evaporate to leave a kind of veneer on the rock surface — the well-known desert varnish.

Attention has recently been paid to the importance of salt crystals. Salts in the form of dry dust may be blown into cracks in the rock. If the salts are moistened by light rain or dew they tend to expand (termed *Salzsprengung* in German, or salt explosion) and rock disintegration results. This type of mechanical weathering is believed to be very important in deserts and especially near dry coasts where sea salt is dried and transported to the land by wind. The quite fantastic rock sculptures observed along the Makran coast of Pakistan, like other *Zeugen*, are almost certainly largely due to this process.

Deflation hollows

These occur in otherwise flat areas of desert. They vary in size from a few metres across to the huge Qattara depression in the Western Egyptian Desert, which is some 32 000 km² in area and reaches a depth of 134 m below sea level. Deflation

Fig. 6.2 *The growth of a deflation hollow*
(a) *An initial surface irregularity or porosity causes a small excess of moisture to be retained that enhances chemical weathering*
(b) *The hollow deepens as weathered products are removed by deflation, until the coarser grains are protected from movement by the wind by the depth of the hollow*
(c) *Chemical weathering as a result of the accumulation of moisture from the desert dews, persists at the base of the sides, which retreat and gradually enlarge the hollow. When the hollow is wide enough the wind is able to remove weathered products again and deepening recommences*

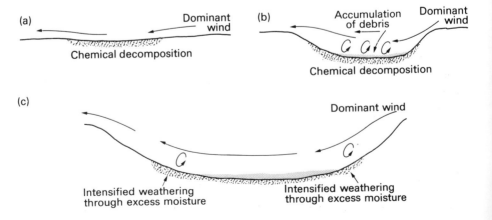

hollows appear to be created through chemical decomposition and transport by wind. An initial irregularity on the surface or greater porosity of the rock causes an accumulation of dew. This in turn promotes an increase in chemical weathering, the products of which, either sand or dust, are removed by the wind, and a hollow formed. As the hollow gets deeper, proportionately more moisture will tend to accumulate, chemical decomposition will be intensified, and the growth of the hollow quicken.

A limiting-point may be reached when the hollow is sufficiently deep to shield the accumulation of comminuted rock from the wind. If this happens deepening will cease. For transport to continue the hollow must be enlarged to enable the wind to move debris from the bottom. The most probable way in which this happens is through weathering at the base of the side walls leading to their gradual recession (Fig. 6.2).

Blown sand

A deposit is defined as sand when the predominant particle size lies between 2·0 mm and 0·06 mm. Sand occupies only a small part of arid areas, which consist mostly of rock and stones, but so large are the deserts that in the Sahara alone sand covers about seven million km². Sand may also form important features in temperate humid areas, for example where coastal dunes form a protective barrier against the sea. Most sand is composed of silica (SiO_2) although it may sometimes be derived from other materials, as in the coral sands of atolls and some South-East Asian beaches, or the basalt sands of Iceland and Hawaii.

When the wind near the ground reaches a critical velocity (which varies with particle size) the sand will begin to move. Sand cannot be sustained by wind in suspension and moves in two different ways, either by creep, or by saltation. Creep is the movement of a layer of sand in continuous contact with the ground in which the grains are rolled along by the influence of the wind. If the velocity of the wind increases sufficiently saltation may take place. This is a type of movement in which individual grains of sand are lifted by the wind and transported forward a short distance in the air before they drop to the ground again. On falling they may displace other particles which are in turn carried a short way in the air, and the process is repeated. When saltation is taking place one may observe myriads of grains of sand moving through the air in the same direction as the wind in a series of hops close to the ground (Fig. 6.3), the height and distance travelled depending upon grain size and wind velocity. It is true that there are records of blown sand grains being found much higher than the normal maximum of 2 m, but almost certainly these were the result of the exceptional desert whirlwind (or dust-devil) which may be of sufficient intensity to carry away a tent or the roof of a house. Although dust-devils may lift sand very high locally they are short-lived phenomena and their direction is random, so they cannot be regarded as transporting agents in the same way as the wind.

Fig. 6.3 *A barkhan dune. Sand moves downwind by saltation and creep. The dune migrates through the shearing and reforming of the slip face.* (a) *In cross-section,* (b) *in plan*

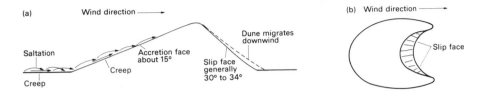

Dunes

In areas where the wind is almost uni-directional, for example in the Egyptian Sand Sea or in southern Peru, barkhan dunes may form. These dunes are shaped like the crescent moon with the horns of the moon pointing downwind. A barkhan may first form wherever an obstruction such as a stone or bush causes blown sand to accumulate. The mound grows as further sand is added. This continues until the height is such that grains of sand from the top are no longer blown clear of the mound but land on its leeward side. Sheltered from the wind the sand grains accumulate and the side steepens until it reaches an angle of about 34°, at which point the slope shears to reform at a slightly lower angle, generally between about 32° and 33°. The leeward face is known as the slip face, and it is by slipping and reforming in this way that the dune migrates downwind (Fig. 6.3). The speed of movement varies in proportion to wind velocity and to the height of the dune, with high dunes tending to advance more slowly. In Egypt during one year a barkhan 20 m high was observed to move nearly 11 m downwind, while one half as high moved nearly 19 m.

As the wind blows over the curved hump of the dune it develops a sideways component. This, and the action of gravity, cause sand to move laterally as well as downwind, causing horns to develop on either side of the slip face, and a typical barkhan is formed. The length of the horns is limited by the protection from the wind afforded by the dune. Once they reach a critical length they experience the full force of the wind and the sand is rapidly removed. This delicately balanced self-regulating mechanism results in the beautifully regular shape of many barkhans. The coarsest sand is normally found near the crest and on the horns, both places where the wind tends to be most effective. The length and width of a barkhan is generally approximately the same, with a height of about one-tenth the width. They may remain quite small, perhaps 2 m or 3 m high and 20 m or 30 m wide, or, under suitable conditions, continue to grow. The ultimate height of a barkhan is controlled by wind strength and sand particle size. The largest barkhans occur in the Egyptian Sand Sea where they may be up to 30 m in height with a width and breadth of around 200 m.

Seif dunes are extraordinary sand features for which there is no sure or generally accepted explanation. They are best developed in southern Iran and consist of single continuous parallel ridges of sand up to 200 m high and in some cases reaching 100 km in length. One explanation suggests that they form under the influence of

Fig. 6.4 *Possible stages in the evolution of a seif dune:*
1. a barkhan is formed under the influence of the dominant wind (S₁),

1. a barkhan is formed under the influence of the dominant wind (S_1),
2. a seasonal change of wind causes one horn to extend,
3. the dominant wind reasserts itself and a slipface forms at the end of the horn (S_2),
4. the sequence is repeated many times until a long straight ridge of sand is left

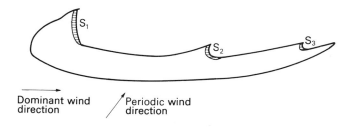

Dominant wind
direction   / Periodic wind
                  direction

a dominant wind from one direction, in areas where there is a regular seasonal change and for short periods of time wind blows from a different direction. Fig. 6.4 shows how this might (possibly) happen. Once the great ridges have formed the dominant wind scours sand from the area between them, which may often be of bare rock. Under the influence of periodic wind change the seif dune may develop a slip face along the side of the ridge. This is only a temporary feature which disappears when the dominant wind reasserts itself.

Consolidation

1. What are the most significant geomorphic processes operating in a warm semi-arid climate? What effects do these have on weathering, erosion, and transport compared with those in temperate humid areas?
2. State in your own words the problems associated with inselbergs, and some tentative possible explanations.
3. Where are the main sand deserts? Describe how sand moves under the action of wind to form dunes.
4. Locate on an outline map areas having a warm semi-arid climate.

Chapter 7

Tides, waves, and beaches

Tides

Tide-producing forces

Tides are produced by the gravitational forces of the sun, moon, and earth, and vary according to the position of all these bodies as they swing round each other. The resultant between gravitational pull and centrifugal effect creates a force which, when it acts in a direction horizontal to the earth's surface, is known as the tractive force. It is this component which creates the tides. The calculation of the tractive force for a given time and place is a mathematical problem which will not be discussed here; what concerns us is the reaction of the sea to it.

The tractive force is at maximum when earth, moon, and sun are in a straight line. This occurs twice every lunar month and results in spring tides when the tide reaches its highest and lowest point. Midway between spring tides occur the neap tides when the tidal range is smallest. High water normally occurs (there are local exceptions) every 12 hours and 26 minutes. This is known as the **semi-diurnal lunar tide**. Tides are important because the tidal range brings the action of the sea to bear on a broad area of land. In nearly tideless seas the effect of the waves is confined to a narrow zone along the shoreline.

Resonance

Other factors also affect the tides. Only the large oceans are big enough to react appreciably to the tractive force. The North Sea is too small to have other than a very small tide of its own, but receives a tidal impulse from the north Atlantic swinging south round the north of Scotland. The water in every sea has a natural period of oscillation. In the North Sea this corresponds closely to the Atlantic tidal impulse which has a cumulative effect on the tides, considerably increasing their amplitude. This is called a state of resonance. The result is that around the area of the Wash in eastern England the range of spring tides averages about 6·7 m and neap tides about 3·4 m. Many small landlocked seas have little tidal variation. For example, in the Mediterranean resonance does not occur because the sea is too deep compared with its size for this to happen. In areas of the south of Mediterranean France the tidal range varies from between 0·18 m at spring tides to only 0·06 m at neap tides.

Amphidromic systems and British tides

Another factor is the effect of the gyration (or spin) of the earth. The earth's spin

Fig. 7.1 *Diagrammatic cross-section of a simple amphidromic system. Note the difference in tidal range between points A and B*

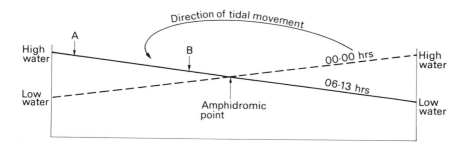

effect is called the Coriolis force by meteorologists dealing with the atmosphere. The effect upon the tidal impulse is much the same as the effect on moving air. It results in a circular anti-clockwise (in the northern hemisphere) movement of the tide. If the ocean basin is of the appropriate length and depth for resonance to occur, a circular movement of high and low water is created known as an amphidromic system. Fig. 7.1 is a diagrammatic section through the centre of a simple amphidromic system in a symmetrical basin. High and low water at a given place are separated by about 6 hours 13 minutes. It will be seen that there is a location at which the level of the sea does not change. This is known as the amphidromic point. The greater the distance from this point the larger will be the tidal range. In Fig 7.1 for example a place located at A would have a tidal range four and a half times as great as one at B.

The three amphidromic systems operating in the North Sea are shown in Fig. 7.2. The centre of the most southerly system is very approximately midway between East Anglia and Holland, and tidal range on both coasts is roughly the same. In the other two systems, however, the amphidromic point is located very much closer to the coast of mainland Europe than to Britain. The range lines show the very great difference in tidal range this causes between, for example, the coast of southern Norway and the east coast of Scotland.

The effect of the shape and depth of ocean basins

The third important factor affecting tides is the shape and depth of the basin in which they occur, because these properties alter the symmetry of amphidromic systems and the natural period of resonance of the basin. The tidal impulse from the Atlantic is modified by the physical characteristics of the North Sea basin to produce the three amphidromic systems described above. In the English Channel and the Irish Sea the small irregularly shaped basins have caused modification to their amphidromic systems. A study of the range lines in the Irish Sea shows the system distorted by friction with an amphidromic point close to the Irish coast, giving a tidal range of less than 1 m south of Dublin but averaging over 8 m on the other side of the sea at Liverpool. An even greater contrast exists in the English Channel with tidal ranges averaging less than 2 m in the area of Bournemouth,

Fig. 7.2 *Co-tidal and range lines for tides in the North Sea, Irish Sea, and English Channel. Co-tidal lines show times of high water. Range lines connect points of equal tidal range*

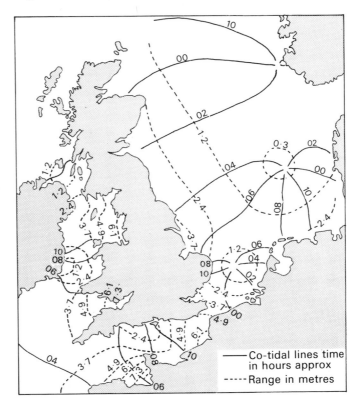

and nearly 14 m, seven times as great, on the opposite side of the Channel near Avranches in France.

In a completely symmetrical amphidromic system the tidal range would increase regularly from the centre in all directions and the range lines would make equally spaced concentric circles centred on the amphidromic point. In nature such a system does not exist because of the effect of irregularities of the basin and friction with the sea floor. However, the system shown in Fig. 7.2 between East Anglia and Holland is very nearly symmetrical. The first circular range line is almost concentric, as are the arcs to the south-west formed between Britain and the continental mainland.

In the English Channel the tides are too complex for a detailed explanation here. It will be seen that the range lines appear to be centred inland on the southern coast of England near Christchurch. In this situation the amphidromic point is said to be degenerate. Notice that in spite of this the range lines, as they move outward from the degenerate (and in this case theoretical) amphidromic point, are spaced remarkably regularly.

Finally, there are places where accidents of size and shape cause exceptional tides. Such an example is the Bay of Fundy with its very great tidal range. Explanation lies in the fact that the natural period of oscillation for the bay is between 11·5 and 13 hours, that is it oscillates very close to the same frequency as the semi-diurnal tide in the main ocean from which it receives its tidal impulse. The consequence is that the effect of the impulse is exaggerated and a very great tidal range results.

Waves

Wave generation

Waves are created by the transfer of energy from wind blowing over the surface of the sea, and are of the type known as progressive waves. Storms are very important in generating waves, which, once formed, can travel very long distances without losing their identity. For example, waves generated by storms in the South Atlantic have been identified in Cornwall. Therefore the characteristics of waves arriving at any coast are not necessarily the result of local weather.

The energy acquired by waves depends on the duration and velocity of the wind and the length of fetch, that is, the distance of open sea in a given direction from a point on the coast. Part of the southern coastline of Britain is open to the Atlantic for several thousand kilometres to the south-west, but only as far as France to the south-east, a distance of 200 km at most and generally less than 100 km. The result is that strong winds from the south-west blowing uninterrupted over a long reach of the open sea can cause waves of high energy to reach the English coast. Whereas a south-easterly wind of whatever strength, because the fetch across the Channel is so short, has insufficient time to create waves of the same high energy, with important consequences for coastal morphology.

When waves are being actively generated under storm conditions they are called sea. As they leave the influence of the strong winds their characteristics change. The sharp crests become rounded, longer, more uniform, and the wavelength increases. Waves modified in this way are called swell. Swell very slowly loses its energy, mainly because of extension over an increasingly wide area of sea. The energy of a wave in deep water is proportional to the product of the wavelength (L) and the square of the height (H), and is given by the formula:

$$E \propto LH^2$$

This formula is included to show how a small increase in wave height will result in a proportionately much larger increase in energy.

Waves in shallow water

Waves are described in terms of height (H), the distance between crest and trough; length (L), the distance between two successive crests; velocity (c), the speed of movement of the crest; period (T), the time taken for a wave to move one wavelength; and steepness (H/L), the ratio between height and length. (H/L can never exceed 1/7, or 0·14, because at this point the wave will break.)

Fig. 7.3 (a) *A progressive wave in deep water. Arrows indicate the direction of movement in each part of the wave form. Arrow heads show the circular orbit of the particles as the wave passes. Velocity is equal in all parts of the orbit* (b) *A progressive wave in shallow water. Particle orbits are now elliptical. The ellipse is compressed towards the sea bed. The layer of water in contact with the sea bed moves backwards and forwards in a straight line.*

The particle velocity is greater in the crest of the wave and reduced as the trough passes. Note the importance of this in causing the movement of material towards the shore, and in the sorting action which results

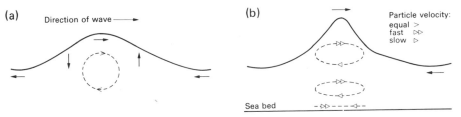

Up to break point the wave form of a progressive wave is transmitted by individual water particles which move backwards and forwards, and up and down in the vertical plane in an open circular orbit to transmit the wave (Fig. 7.3a). The motion of the water particles extends downwards on a decreasing scale to a depth equivalent to about one wavelength. Under normal conditions there is a slight mass movement of water in the direction the wind is blowing, called the mass transport velocity. It is now recognized that this may be greatly increased by strong storm winds.

As the wave approaches shallow water its characteristics begin to change. Shallow water may be defined as when the depth approximates to half the wavelength. As the water gets shallower the circular movement of the water particles becomes elliptical, with the long axis horizontal to the bottom. The closer to the bed of the sea the narrower will be the ellipse (Fig. 7.3b).

Wave forms generated in different places and having different characteristics in terms of height, wavelength, and period, can pass through each other undisturbed. This can often be seen quite clearly from the top of high cliffs; or it may be demonstrated by dropping two stones into still water, where the two sets of ripples (i.e. the wave forms) will be seen to pass through each other and remain intact. If different wave forms approach a beach from the same direction they may become superimposed in a complex manner and result in a succession of waves of varying sizes breaking upon the beach. A component with a comparatively long wavelength can cause a regular succession of high waves amongst a series of smaller ones. This may help to explain the popular myth that every seventh wave is a big one.

When the water is very shallow the velocity of sea waves is controlled by the depth (exactly as it is in streams) and is given by:

$$c = \sqrt{gD}$$

where g is the acceleration of gravity and D is the mean depth of water. The waves will therefore slow down as the water shallows, causing a corresponding reduction in wavelength. In very shallow water as the value of the ratio of depth to length (the D/L ratio) passes 0·06, the height of waves with a long wavelength increases rapidly up to break point. Waves with a shorter wavelength are less affected though some increase takes place. The wave also steepens with the crest becoming narrower. The velocity of the orbiting particles begins to change, with the movement forward in the crest of the wave accelerating, and that in the trough slowing down (Fig. 7.3b). The speed of the water in contact with the sea bed is also faster forwards than backwards. This differential movement of water near the sea bed as each wave passes is very important because the faster forward thrust results in a continuous movement of material towards the shore. Waves break when the depth of water is about one and a third wave height at break point. As the water gets shallower the wave gets progressively higher and steeper with the forward speed of the orbiting particles in the wave crest continuing to accelerate. When their velocity exceeds that of the wave form the wave breaks. Additionally, as the forward velocity of the particles increases so does the size of their orbit. At the same time the shortening wavelength reduces the amount of water available for the wave. At break point the amount of water becomes insufficient to complete the particle orbits.

Wave refraction

In shallow water the direction of movement of a wave is affected by submarine relief, the wave crestlines tending to become parallel to the submarine contours. If we consider an indented coastline the submarine contours often follow the same general pattern as those above sea level. As the waves approach the shore their velocity is reduced as the depth decreases. This does not take place evenly along the length of the wave. Because the water shallows more quickly off the headlands (Fig. 7.4a), the movement shorewards is here slowed down, while the remainder of the wave, still in the deep water of the bay, is unaffected until the shallows are reached. The result is that the line of the wave tends to become parallel to the shore.

The effect of this refraction is to alter the energy distribution along the waves. In deep water this is the same for equal unit lengths all along the crestline, but in shallow water as parts of the wave tend to converge and parts to diverge so energy per unit length is concentrated in areas of convergence and reduced in areas of divergence. This can be seen from the wave orthogonals shown in Fig. 7.4a. (Wave orthogonals are imaginary lines drawn at right angles to the line of the crest of the wave.) The energy of a unit length of the wave at A is expended over a much longer length of shoreline at A_1, whereas the energy of a similar unit length of wave at B is concentrated into a much shorter length of shoreline at B_1. The result is to concentrate the destructive effect of the sea upon the headlands. This may also cause a slight local raising of the sea level and create a longshore current (and movement of material) from the headland into the bay.

Along straight shorelines, if the submarine relief is regular, the only change in waves advancing parallel to the coast is a shortening of the wavelength. If waves approach obliquely, as in Fig. 7.4b, the part of the wave nearest to the shore will be retarded and the crestlines will bend round towards the beach. Even on a straight shoreline submarine relief affects the wave orthogonals, and thus the distribution of energy reaching the shore; a submarine valley will cause divergent and a submarine ridge convergent orthogonals.

Fig. 7.4 *The effect in shallow water of submarine relief on wave refraction.*
(a) *An indented coastline. (A common form of coast today, following the eustatic rise in sea level, mainly due to the melting of the Pleistocene ice sheets.) Note the concentration of wave energy on the headlands*
(b) *A straight shoreline with regular submarine relief*

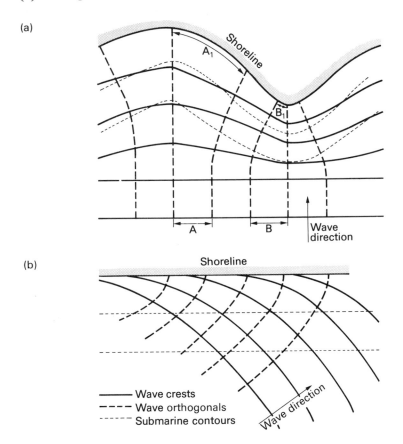

(a)

(b)

Dominant and prevalent waves

Care must be taken to distinguish between these two terms. Prevalent waves are those which come most frequently from a given direction. Dominant waves are those which have most effect on the coast in terms of erosion and deposition. For example, along most of the coast of south-west England dominant and prevalent waves come from the south-west because that is the direction of longest fetch, and also the direction of the prevailing winds. In detail the direction of dominant waves varies considerably from place to place, depending on the alignment of the beach and the existence of headlands or other forms of protection.

Constructive waves

From its generation in deep water right up to break point we have seen that a sea wave consists of orbiting particles. For this reason it is called a wave of oscillation. At break point the orbits are broken and thereafter the water rushes forward in a mass. The confused mass of water which moves up the beach is called swash. The water running down the beach is backwash. The swash from a constructive wave has the energy of the breaking wave behind it as it pushes up the beach. As it (or most of it) returns as backwash there is only gravity to overcome friction from the beach material. The result is a net movement of material up the beach. Constructive waves are the result of swell. They tend to be flat, not very steep, and have a long wavelength relative to their height, with an average period of about 7 to 10 seconds.

Destructive waves

Destructive action is characteristic of storm waves which are steeper, higher, and with a much shorter wavelength than constructive waves. Onshore storm winds cause the mass transport of surface water shorewards, which in turn creates a counter-current flowing seawards along the bottom. (Or the excess water may sometimes break back seawards in depth at a particular point, causing a rip current.) The effect, if the shore is tidal, is to flatten the lower part of the beach by combing a mass of material down towards the sea, and to leave a steeper profile in the area of high-water mark.

Waves, cliffs, and sea defences

There is also the special case where waves are able to attack directly rocks, cliffs, or man-made sea defences. Waves breaking on coasts exposed to a long fetch, such as in the west or south-west of Ireland, can release great energy. Pressure exerted by large storm waves may be as much as 30 tonnes weight or more per square metre. Whether this happens depends mainly on the existence of a pocket of air trapped between the breaking wave and the sea wall or cliff (Fig. 7.5). The depth of water is critical because if it is too deep the wave will not break but simply be reflected back, or if too shallow the wave breaks before arriving at wall or cliff. In either case, pressure is greatly reduced and potential erosion negligible.

If the sea wall or cliff is intact the release of energy, even of the order quoted above, may do little or no damage, but should there be a fault in the concrete, or a joint plane in the cliff, then the hydrostatic pressure developed through the pocket of air may be sufficient to cause very considerable damage. Once erosion has begun, succeeding waves are able to exploit and enlarge any weakness.

Fig. 7.5 *A wave in shallow water breaking against a sea wall. Great hydrostatic pressure may be generated if a pocket of air is trapped as illustrated. If the water were deeper and the wave failed to break, or shallower and the wave broke before reaching the wall the resulting pressure would be negligible*

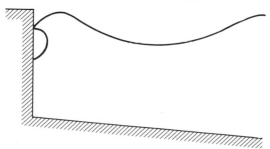

Wave-cut platforms

If waves in a tidal sea are able directly to attack the land, a wave-cut platform may result, if sea level remains unchanged for a sufficient period of time. The length of time required will depend upon the resistance of the rock. The platforms tend to be very slightly concave in profile, with the gradient depending mainly on the average characteristics of the waves. The slope has been found to vary from between about 1 in 50 to 1 in 90 or 1 in 100. Once the equilibrium gradient has been attained no further erosion will take place unless the characteristics of the waves change or sea level rises. With an unaltered sea level the extent of the platform depends upon the tidal range. Between low and high water the rock is exposed to quarrying by the waves. If the debris is subsequently removed by the sea and no beach forms a wave-cut bench will be initiated. Below water, as in the tidal section, the rock will undergo the normal processes of chemical decomposition, possibly including solution. It may also suffer abrasion if a little sand or gravel is present. If this is in the form of a complete layer, thick enough to prevent movement by the waves of the grains in contact with the rock (the depth of disturbance in sand has been shown to be about 1 cm for every 30 cm of wave height), it will act as a protective cover. But if the grains are sufficiently large — a minimum diameter of 1 mm has been suggested — and are rolled backwards and forwards on rock, the abrasion can cause significant erosion. It had been held in the past that, in theory, waves might move grains of this size down to a depth of 90 or even 180 m. Today it is believed the depth limit to effective movement on the sea bed is between 10 and 15 m.

Provided sea level remains constant the maximum possible width of a wave-cut platform is constrained by the tidal range and the equilibrium gradient which develops. Even if we assume an extremely large tidal range of 10 m the width of the platform is limited to 500 m for a gradient of 1 in 50, and 1000 m for a gradient of 1 in 100. It might be argued that submarine erosion can also take place down to a depth of 10 m and that the platform might theoretically be extended by this means a further 500 or 1000 m respectively beyond low-water mark. With a tidal range of 10 m this limits the maximum width of the platform to 2000 m.

If we wish to consider marine planation as an explanation for an erosion surface wider than 2 km we have also to assume a very long and slow rise in sea level, so that new land is steadily being presented to the sea at exactly the rate at which material is being removed. For an area of marine planation several kilometres in width the critical rate of base level change would have to persist for a very long period of time indeed, and the interpretation of such areas in these terms should be viewed with caution.

Beaches

Longshore drift

The action of the sea upon the land may erode material that is small enough for the waves to move, or result in larger rocks that over a period of time are slowly reduced to shingle or sand. Rivers are also continually bringing sediments down to the sea. All of which may then be moved along the coast by the process known as beach

Fig. 7.6 *Beach drifting: this is caused when waves break obliquely on to a beach*

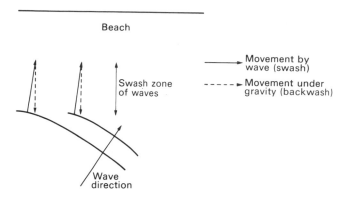

drifting (Fig. 7.6) and by longshore currents. The most important transporting agent is probably the current which develops below the break point of the waves in the direction of wave travel. Other currents may be present, but the process is complex and subject to much local variation and is not considered further here. The movement of all material along the shore, by whatever process, is termed longshore drift. If the sea at a given place is able to remove more than the material received there will be no beach and, unless measures of artificial protection are introduced, the coast will be actively eroded. In areas where more material is received than can be removed, a beach will form. If both are in balance an equilibrium beach will result. Longshore drift is sometimes interrupted by headlands. If the water at the seaward end is 10 m deep or more the waves are unable to move much beach material round it unless accumulation reduces the depth of water. For some beaches a purely local equilibrium exists, especially bay-head beaches. That of Slapton Lea in Devon is a good example, where local drift is slight and nicely in balance in each direction between two headlands.

We have seen (above) how wave refraction on an indented coast tends to concentrate wave energy, and erosion, on headlands. Under wave attack these will eventually disappear and the coastline straighten, often with a continuous beach running along it. The process of longshore drift ensures that beaches tend eventually to become perpendicular to the direction of onset of the dominant waves. Once this stage is reached longshore drift, under the influence of the dominant waves, remains important only to correct small movements of material that may occur during seasonal or local variations in the direction of wave incidence.

Longshore drift does not always lead to a straightening of the coastline. There is not space here to discuss the formation of tombolas and the many kinds of spit and other features that may result. But some mention must be made of the cuspate foreland of Dungeness, because it is an interesting example of the interaction of wave energy and longshore movement.

Dungeness lies on the south coast of Britain at the eastern end of the English Channel between Rye and Hythe. 2000 years ago sea level was probably nearly 2 m lower than today. At this time the shoreline lay about 16 km inland from the present ness (Fig. 7.7). The dominant waves in this area are from the south-west

Fig. 7.7 *Dungeness*

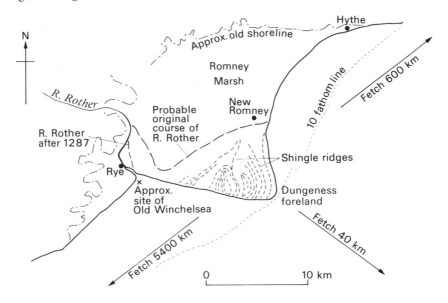

and longshore drift is mainly from west to east. It will be seen that although by far
the longest fetch, 5400 km, is from the south-west, there is a fetch to the north-east
of some 600 km. So while there is a general movement of beach material eastwards
there is sometimes a return movement westwards under the influence of easterly
winds.

The foreland probably started as a long spit from the coast west of Rye and
eventually stretching to Hythe across what was then the mouth of a bay. 1287 was
a year of great storms during which Old Winchelsea was almost entirely destroyed
by the sea and the River Rother broke through the shingle banks near Rye and
abandoned its old outlet to the sea near New Romney. The rapid accumulation of
shingle to form Dungeness dates from this period. Under the influence of waves
from the south-west, assisted sometimes by waves from the north-east, the foreland
grew outwards into the Channel. As it grew the beach tended to build up at right
angles to the line of advance of the dominant waves. The sharpness of the point is
maintained because the distance to the coast of France is only 40 km — insufficient
fetch for the strongest south-easterly winds to generate waves of sufficient energy
to destroy it. The limiting factor to the extension of the foreland seawards is depth.
It will be observed how close to the point the 10–fathom line runs.

The shingle ridges, which are clearly visible on the ground, are the storm beaches
of successive shorelines. As shingle accumulated the distance between each successive
storm beach and the sea was increased and a new storm beach formed. There is
reason to believe from historical sources, and from the pattern of the present ridges,
that Dungeness originally began to form a little to the west of its present position,
and under the influence of the dominant waves then moved, and is probably still
moving, towards the east.

Shingle beaches

On the geological time-scale shingle beaches are features of considerable rarity, because unless the supply to the beach is continuously renewed the abrasive action of the waves will reduce the pebbles to sand. Shingle is comparatively common today as a result of the unusual conditions of the Pleistocene which provided the sea around our coasts with extensive deposits containing gravels and rock fragments. The rise in base level after the final glaciation carried the sea over these deposits. This may possibly be the origin of the shingle of the Chesil Bank, which is really a relict feature as the addition of new shingle has long ceased.

Beaches are commonly divided for convenience into three sections – the backshore, the foreshore, and the nearshore. The backshore is the area above spring tide high-water line. The foreshore is the area between the spring tide low- and high-water line. The nearshore is the sea floor below low water influenced by the waves, that is, roughly to the line where the sea is 10 m deep at low water.

The equilibrium gradient of a beach depends upon the size of the material of which it is made and the characteristics of the dominant waves. The foreshore of shingle beaches is generally steep, because water percolates easily through shingle, reducing the amount of backwash from a constructive wave and making the action of the swash more effective. Also the maximum angle at which dry unconsolidated material will stand depends upon particle size. If the pebbles are large, as at the eastern end of the Chesil Bank, parts of the beach may be very steep indeed, with gradients of up to 1 in 2. At neap tides a typical shingle beach consists of three distinct ridges (Fig. 7.8): a storm beach, a ridge associated with spring tides, and a third associated with neap tides. There may also be minor ridges caused by other tides and especially the high spring tides of the spring and autumn equinox.

The storm beach is found above high-water mark and is out of reach of the sea under normal conditions. Normally shingle cannot be carried in suspension by water, but under storm conditions the pressures created by waves breaking on the beach may be sufficient to fling jets of water into the air, and with them may be carried quite large stones. Some of these may land above the highest tide level and accumulations of these stones form a storm beach. Destructive storm waves may thus have this constructive aspect on shingle. The other two main ridges are caused by the

Fig. 7.8 *Profile of a shingle beach at the time of neap tides. The ridges mark accumulations of shingle at spring and neap high water. The storm beach exists above the highest tides formed by the stones flung up under storm conditions. The small ridge formed by high water at the spring and autumn equinoxes may often not be identifiable*

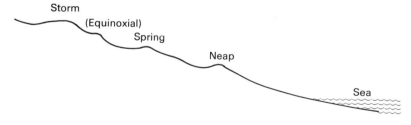

Fig. 7.9 *Concentration of the swash zone of breaking waves due to the steeper angle of a shingle beach*

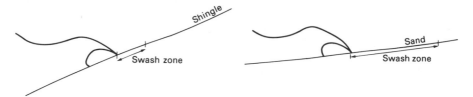

accumulation of material along the high-water mark of spring and neap tides. The neap tide ridge disappears as high water rises towards the spring tide level. A shingle beach reacts more quickly to wave action than sand, as the steep angle concentrates the energy of the breaking wave expended on the beach by decreasing the area of the swash zone (Fig. 7.9). Shingle beaches are therefore more mobile than sand beaches.

Sand beaches
Sand beaches differ from shingle beaches in a number of respects, the most obvious being that they are generally very much flatter. Because of the small size of the grains of which a sand beach is composed, there is comparatively little percolation especially once the surface is wet, and consequently virtually all the swash returns down the beach as backwash carrying with it a high proportion of the sand previously transported upward. This is the main factor in maintaining a profile of low gradient.

For example, Blackpool beach has a gradient of about 1:160, and in northern France there are silt beaches with gradients as low as 1:400. We have seen (page 66) that wet sand coheres due to chemical bonds formed by the moisture. When in this condition sand can, and sometimes does, maintain quite a steep profile, but this is rarely found on beaches, except for the banks of some runnels if these develop on the foreshore, or temporarily near high-water line after a storm. The gentle gradient of the sand beach causes the energy from the swash of the waves to be extended over a wider area (Fig. 7.9) and therefore sand beaches react less quickly to wave action.

Sand beaches do not exhibit the marked tidal ridges of shingle beaches; they normally exist, although because the change in gradient is slight they may sometimes be difficult to detect. Sand is also easily moved by the wind and a slight ridge left by spring high tides would tend to be removed before the next spring high. Sand may be carried in suspension by the sea, but it cannot be thrown through the air like shingle, and so a storm beach rarely develops. Where it does — and there is an example at Gibraltar Point on the Lincolnshire coast — the sand of which it is composed is carried in suspension in droplets of seawater flung by storm waves above the level of high spring tides. A ridge is frequently found along the line of spring high water (additionally sometimes along the line of the spring and autumn equinoctial high water) containing a large proportion of bigger than average sand grains. This ridge is generally the result of sand deposited by constructive waves at spring high tides and subsequently acted upon by the wind which removes the smaller grains.

Wind can be an important agent in the transport of dry sand. Even when the beach sand is saturated with water individual surface grains dry off remarkably quickly and large amounts can be transported by saltation. The process is important because wind-blown sand may result in the formation of dunes which, unlike desert dunes, may be 'fixed' naturally or artificially by vegetation to form a protection for the coastline against the sea.

Changes in sea level

These are important for two reasons. Firstly because of the influence base level change has upon streams and the consequent effect on landforms; and secondly because with a rising sea level new land is continually being presented to the sea, resulting in new marine features of erosion and deposition. Changing sea level may be a local phenomenon or it may be eustatic (worldwide). Local changes may be caused by downwarping of the earth's crust to give an apparently rising level, as in south-east England and in Holland. Or the sea level may be apparently falling as in north-west Scotland (and even more so in Scandinavia) due to isostatic readjustment caused by the removal of the weight of the Pleistocene ice.

Eustatic changes are very complex and are due to a number of causes. The most rapid fluctuations are through the transfer of water from the oceans to the land to form ice sheets in periods of glaciation and its subsequent return to the sea, in whole or in part, during interglacials. There is evidence that the sea was at least 100 m below the present level and possibly much lower at the end of the last major glaciation about 20 000 years ago. Equally important in the long term are changes in the capacity of the ocean basins. It has been estimated that the uplift of the southern mid-Atlantic ridge in the late Miocene could have resulted in a eustatic rise of sea level of up to 500 m. Another factor causing sea level to rise, although probably rather less important than the creation of submarine mountains, is the deposition in the sea by the rivers of material eroded from the land.

Movements of sea level at any place are the result of some or all of these factors. This is well demonstrated in Britain where parts of the north-west are rising, due to isostatic readjustment, at rates of up to +4·0 mm a year, causing the raised beaches of western Scotland. The south and east are slowly sinking, reaching a maximum measured at Newlyn in Cornwall of −2·3 mm a year, as part of a synclinal development of ancient origin in the southern North Sea. Human problems arising from this slow but steady rise in sea level are being felt increasingly acutely in Holland, and to a lesser extent in London, the Thames estuary, and the Lincolnshire fens. .

Storm surges

Atmospheric disturbances may considerably affect the height of tides, especially in enclosed areas like the North Sea. If atmosphere pressure were evenly distributed over a still and tideless sea the surface of the water would be flat. If the pressure distribution changed to give one area of high and one of low pressure the level of water would respond by becoming depressed in the area of high pressure and elevated in the area of low pressure. In practice a fall in pressure of 56 millibars causes sea level to rise 0·5 m, and a corresponding rise depresses sea level by a similar amount. Air pressure is therefore one factor in modifying predicted tide levels.

The North Sea is an area where tides occasionally very much exceed the calculated level. These abnormally high tides are known as tidal surges, and can be very danger-ous for those living below mean high-water mark, as many do in the English fenlands and in Holland. There is historical evidence that surges have caused great damage over past centuries. More recently, in addition to a number of small surges, there was a particularly severe one in 1897, followed by another in 1953.

Surges may be caused simply by an increase in the tidal impulse from the outside ocean, but a really disastrous rise in sea level seems to occur only when certain meteorological conditions are fulfilled within the area of the North Sea itself. The surge which took place between 30 January and 1 February 1953 provides a good example of the factors involved in raising the level of high water. On 29 January a depression of considerable intensity was situated north-west of Scotland, while at the same time a ridge of high pressure was developing west of Ireland. Normally a depression in this area moves away eastwards over Scandinavia. On this occasion it moved south-east across the North Sea, and eventually into Germany. Fig. 7.10 shows the synoptic situation at 12.00 hours on 31 January. At this time there was a difference in pressure of 66 millibars between the high off Ireland and the low midway between Scotland and southern Norway. The steep pressure gradient, demonstrated by the compression of the isobars, caused very strong northerly winds to blow down the east coast of Britain, that is, from the direction of maxi-mum fetch. The results were storm waves up to 6 m high of very great energy, and the mass transport of water southwards due to the drag effect of the wind on the sea's surface. Of great importance in increasing the size of the surge was the speed of movement of the depression. Since the speed of a low pressure system in a small sea is critical in relation to the moving wave of high water of the amphidromic

Fig. 7.10 *Simplified synoptic chart for the area around Britain at 12.00 hours on 31 January 1953. Arrows indicate the track of the low pressure centre*

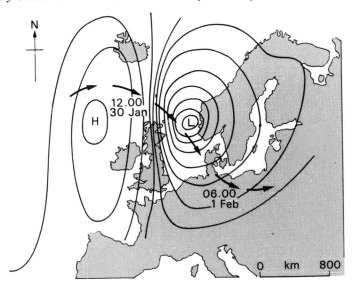

system, the amplitude of the normal tidal wave may be considerably increased. To take an extreme case, a low pressure centre of 947 mb (the lowest recorded) moving over water 50 fathoms deep at 50 knots could increase the height of predicted high water by 2·5 m.

It is difficult precisely to assess the contribution each of these factors made to the surge of 1953. But the result was disastrous. On the Lincolnshire coast high water rose 2·4 m above the predicted level. At Dover it was about 2 m higher than predicted, and in Holland 3·1 m higher. (These figures do not take into account the height of the waves.) The surge was especially dangerous as it occurred at a time of spring tides, when high-water levels were to be expected anyway. Fortunately, the maximum disturbance differed from the time of high water by about two hours, considerably diminishing the elevation of the sea. Had these coincided exactly, the sea level would have been considerably higher and the disaster much greater. As it was storm waves were brought high up on to the sea defences which were breached in many places. In addition to much material damage, 264 people were drowned in Britain and over 1350 in Holland. Since 1953 the sea defences in eastern Britain have been strengthened, but if on some future occasion similar or more intense meteorological conditions occur at a time of very high spring tides, and if the movement of the disturbance coincides with that of high tide, the flooding of large areas is inevitable. In human terms, because of high population densities, the hazard is particularly dangerous in parts of London and the Thames estuary.

Consolidation

1. Describe briefly the tidal systems of the North Sea. What are the social and economic implications? What causes a storm surge, and why is it potentially so dangerous in the North Sea?
2. Describe and explain the principal differences in the profiles of sand and shingle beaches.
3. Demonstrate diagrammatically the limits to the width of a wave-cut platform, assuming no change in sea level.
4. What is wave refraction? How does this affect the coastline?
5. How do waves move material shorewards in the nearshore zone? How do storm waves break up rocks and sea defences, and comb back material from the beach?
6. In non-mathematical terms explain why a wave breaks.

Examination questions

This sample of advanced level questions has been selected from six examining boards to represent the kind of questions for which this book has a special contribution to make. Many other questions, and from other boards, might have been included. As far as possible the selection has been made to avoid overlap in the material required for answers.

1. Why should physical geographers wish to know the levels of water tables? (Oxford and Cambridge, 1975)
2. Give an explanatory account of the processes by which weathered material is produced and transported downslope in two of the following climatic regions: (i) humid cool temperate; (ii) tundra or sub-arctic; (iii) hot desert. (Joint Matriculation Board, 1977)
3. How far is it true to say that different climates produce different kinds of rock weathering? (Cambridge, 1977)
4. What are the main factors influencing the form of slopes? (Cambridge, 1977)
5. The form of river valleys is better explained by reference to rate of river erosion, geological structure, and climate than to stage of development. Discuss. (Cambridge, 1977)
6. How do glaciers erode their valleys? (Cambridge, 1977)
7. Discuss the ways in which wave action modifies the form of beaches, both in profile and in plan. (Cambridge, 1977)
8. Chemical weathering is an important process only in limestone areas. Comment critically on this statement. (Cambridge, 1976)
9. Discuss the form and origin of depositional features in hot deserts. (Cambridge, 1976)
10. Before the Ice Age, fluvial erosion led to the development of a maturely dissected landscape in an upland area in south-west England. The area was not glaciated during the Ice Age but the lower reaches of the rivers were incised and slopes of up to 40° developed on the valley sides.

    Describe the processes of weathering and mass wasting operating on the slopes and leading to their subsequent modification (i) under the periglacial conditions during and immediately after the Ice Age; (ii) under present-day climatic conditions. (Joint Matriculation Board, 1976)
11. Assess the relative importance of wind and water in the development of landform in semi-arid areas. (Joint Matriculation Board, 1977)
12. An upland area experiences a warm temperate climate. It has a thick regolith and receives over 1000 mm of rain per year in the form of torrential downpours

of short duration. Explain the changes in the landscape morphology and river profiles which would follow the deforestation of the area. (Joint Matriculation Board, 1977)

13. Write an essay on one of the following: man as a geomorphological agent; stream hydraulics. (London, 1976)

14. Write an essay on one of the following: the concept of dynamic equilibrium; hydrology as a part of physical geography. (London, 1975)

15. What do you understand by the term 'periglacial processes'? With what distinctive landforms are such processes associated? (Oxford, 1977)

16. Examine the difference between glaciation and periglaciation. (London, 1977)

17. Either: Examine the factors which influence the runoff characteristics of a drainage basin.
Or: Describe how you would attempt to study the flow and channel characteristics of a small river. (London, 1977)

18. Compare the denudation processes at work in hot derserts with those operating in polar deserts. (London, 1977)

19. Examine the reasons why slope profiles vary greatly. (London, 1976)

20. Examine the part played by water in the weathering of rock and the mass movement of regolith. (London, 1977)

21. Either: Explain what is meant by 'the theory of dynamic equilibrium'. To what extent does it help in our understanding of landforms?
Or: Discuss the value of measurement in geomorphology. (London, 1976)

22. Consider the various geomorphological processes at work within a river basin in a region of temperate humid climate. (Southern Universities, 1974)

23. 'A study of process is the necessary key to an understanding of slopes'. Discuss. (Oxford, 1975)

24. Consider the geomorphological processes at work in areas of tropical arid climates and describe the landforms which result from these processes. (Oxford, 1975)

25. Either: Assess the importance of longshore drift in the formation of constructional features along the coast.
Or: 'Wind action has produced significant landforms only in the hot, arid areas of the world'. Discuss. (London, 1977)

26. Assess the relative importance of wind and water in the development of landforms in arid areas. (Oxford, 1976)

27. Write a clearly argued essay on the role of water in the weathering of major landforms. (London, 1975)

28. How do rivers erode their valleys? Choose actual examples to illustrate your answer. (London, 1975)

29. Discuss the different criteria that might be used in a classification of slopes. (London, 1975)

30. Explain the relationship between mass wasting and slope development. (Joint Matriculation Board, 1977)

Sources

Works referred to in the text
† Chorley, R.J., and Kennedy, B.A., *Physical Geography: a systems approach* (Prentice-Hall, 1971)
† Doornkamp, J.C., and King, C.A.M., *Numerical Analysis in Geomorphology* (Arnold, 1971)
† Embleton, C., and King, C.A.M., *Glacial Geomorphology* (Arnold, 1975)
† Embleton, C., and King, C.A.M., *Periglacial Geomorphology* (Arnold, 1975)
† Gregory, K.J., and Walling, D.E., *Drainage Basin Form and Process* (Arnold, 1973)
† Leopold, L.B., Wolman, M.G., and Miller, J.P., *Fluvial Processes in Geomorphology* (Freeman, 1964)
* Morisawa, M., *Streams* (McGraw-Hill, 1968)
* Schumm, S.A., and Lichty, R.W., 'Time, Space, and Causality in Geomorphology', *American Journal of Science*, vol. 263 (1965)
* Small, R.J., *The Study of Landforms* (Cambridge University Press, 1970)
* Sparks, B.W., *Geomorphology* (Longman, 1973)
* Weyman, D.R., *Runoff Processes and Streamflow Modelling* (O.U.P., 1975)
* Young, A., *Slopes* (Longman, 1975)

Additional bibliography
† Chorley, R.J., *Introduction to Fluvial Processes* (Methuen, 1971)
† Chorley, R.J., (ed.), *Water, Earth, and Man* (Methuen, 1969)
* Cooke, R.U., and Doornkamp, J.C., *Geomorphology in Environmental Management* (Oxford University Press, 1974)
* Hanwell, J.D., and Newson, M.D., *Techniques in Physical Geography* (Macmillan, 1973)
† King, C.A.M., *Beaches and Coasts* (Arnold, 1972)
† King, C.A.M., *Techniques in Geomorphology* (Arnold, 1966)
* Newson, M.D., *Flooding and Flood Hazard in the U.K.* (O.U.P., 1975)
† Pitty, A.F., *Introduction to Geomorphology* (Methuen, 1971)
† Russell, R.C.H., and MacMillan, D.H., *Waves and Tides* (Hutchinson, 1952)
* Whalley, B.W., *Properties of Materials and Geomorphological Explanation* (O.U.P., 1976)

* Good clear outlines of subject under discussion. Not too advanced for sixth-form and college students, although parts may be considered demanding.

† Rather more advanced texts, but still worth referring to and containing sections that may be read easily.

Index

Acknowledgements

We gratefully acknowledge permission to use copyright material, the sources of which are as follows:

From S.A. Schumm and R.W. Lichty, *American Journal of Science*, Vol. 263 (1965), Fig. 1.3; from R.J. Chorley and B.A. Kennedy, *Physical Geography: a systems approach*, Prentic-Hall Inc., Fig. 1.4; from R. Hammond and P.S. McCullagh, *Quantitative Techniques in Geography*, Oxford University Press, Figs. 2.2a, 2.4, 3.13; from M. Morisawa, *Streams*, McGraw-Hill, Fig. 3.3; from R.U. Cooke and J.C. Doornkamp, *Geomorphology in Environmental Management*, Oxford University Press, Fig. 3.6; from J.C. Doornkamp and C.A.M. King, *Numerical Analysis in Geomorphology*, Edward Arnold, Figs. 4.2, 4.10; from A. Young, *Slopes*, Longman, Fig. 4.7; from J.B. Dalrymple, R.J. Blong, and A.J. Conacher, *Zeitschrift fur Geomorphologie*, Vol. 12 (1968), Fig. 4.8; from R.J. Small (after M.J. Clark), *The Study of Landforms*, Cambridge University Press, Fig. 4.9b; from P.J. Williams (after G. Beskow), *Geographical Journal*, Vol. 123 (1957), Fig. 5.2; from P. James, *Geographical Magazine* (September 1972), Fig. 5.3; from A.H. Lachenbruch, Geological Society of America Special Paper 70, Fig. 5.4; from J.R. Mackay, *Geographical Bulletin* (1962), Fig. 5.5; from J. Tricart and A. Cailleux, *Le Modèle glaciaire et nival*, Société d'Edition d'Enseignement Supérieur, Fig. 5.8; from B.A. Chart No. 301 (5058) with the sanction of the Controller, H. M. Stationery Office and of the Hydrographer of the Navy, Fig. 7.2.

Appreciation is due to the following examining boards for permission to use the examination questions which appear on pp. 124–5: Joint Matriculation Board; Oxford and Cambridge Schools Examination Board; Oxford Delegacy of Local Examinations; Southern Universities Joint Board; University of Cambridge Local Examinations Syndicate; University of London University Entrance and School Examinations Council.

We should also like to thank Peter James, University of Liverpool, for supplying Photos 4.1, 5.3, and 5.4.